The History of Aquaculture

The History of Aquaculture

Colin E. Nash

WILEY-BLACKWELL

A John Wiley & Sons, Ltd., Publication

Blackwell Publishing was acquired by John Wiley & Sons in February 2007. Blackwell's publishing program has been merged with Wiley's global Scientific, Technical, and Medical business to form Wiley-Blackwell.

Editorial Office
2121 State Avenue, Ames, Iowa 50014-8300, USA

For details of our global editorial offices, for customer services, and for information about how to apply for permission to reuse the copyright material in this book, please see our Website at www.wiley.com/wiley-blackwell.

Library of Congress Cataloging-in-Publication Data

Nash, Colin E.
  The history of aquaculture / Colin E. Nash.
    p. cm.
  Includes bibliographical references and index.
  ISBN 978-0-8138-2163-4 (hardcover : alk. paper)
  1. Aquaculture–History. I. Title.
    SH21.N37 2011
    639.809–dc22

                        2010030978

A catalog record for this book is available from the U.S. Library of Congress.

Set in 10/12.5 pt Sabon by Aptara® Inc., New Delhi, India

1 2011

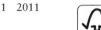

Dedication

I would like to dedicate this book to the memory of Fred Hickling, who first introduced me to the history of aquaculture many years ago.

And to my wife, Patricia Bainbridge Nash, for patiently sharing and supporting my life's work in the modern aquaculture field.

Contents

Abstract

From the earliest prehistory, seafood was an accessible and nutritious resource. Starting with hunting and gathering fresh and saltwater life, husbandry was recorded in China as early as 4000 years ago.

Via a lifetime of personal research and experience, the author relates an intriguing story of the development of culturing fish, seaweeds, shellfish, and other organisms. The History of Aquaculture draws on the literature and records spanning millennia. Aquaculture is traced from its origins in China via Roman vivariae piscine (fish ponds) through technical and scientific advances to the twentieth century's expansive growth and globalization.

Today, aquaculture complements wild-catch fishing as a sustainable source of high-quality protein. Although aquaculture in the twenty-first century starts with many new species under culture, it faces many challenges. The future of global aquaculture will depend not on further technological development, but rather on public demand, markets, and commitment to its further success.

Acknowledgments

The conception of *The History of Aquaculture* has an appropriately historical but personal explanation of its own. In 1965, when presenting my very first paper on marine fish farming to the Challenger Society in London, I was fortunate to be on a panel chaired by the eminent Fred Hickling, who had just retired from the Colonial Office after decades of service overseas predominantly concerned with developing freshwater fish farming. He had returned to research at the Plymouth Laboratory and in the Fish Section in the bowels of the Natural History Museum in London. Hickling was genuinely excited by the rejuvenation of interest in marine fish farming in the British Isles by the White Fish Authority and consequently, was eager to help me with the thermal-rearing studies I was conducting in power plants in South Wales and Scotland by collecting juveniles of flatfish and mullets in the estuaries around Plymouth and the Welsh coast. It was on such trips that I first learned that fish farming had a very long and fascinating history, while he was using his newly found time to travel around Europe to research the records of the medieval monasteries and estates to learn more about their old "stew ponds."

Because I am a person who likes to know the reason why things are as they are and an inveterate collector of old things, it was quite natural for me to follow Hickling's example. I began to gather pieces of information here and there, which to me were interesting aspects of the history of my chosen field, and to simply squirrel them away. Fortunately, for the next forty years, my work enabled me to spend time with many of the pioneers of the modern field and to visit most of the key marine research stations that had played a role, albeit perhaps small, in its historical evolution. Consequently, my collection of first- and second-hand research information and vignettes of aquaculture's development grew until it could be collated into this book.

Burr Steinbach, for many years my mentor and colleague at the Oceanic Institute in Hawaii and former director of the Marine Laboratory at Woods Hole, said that you could usually judge the work of a research center by spending time in the library and talking with the librarian. I came to know the truth in those words many times as I spent long hours searching and verifying information from among the archives and shelves of some excellent libraries of many facilities all over the world. However, I would particularly like to recognize the special help given to me in these quests by Flora McLeod, the librarian of the Marine Biological Station, Port Erin, Isle of Man; Clare Cuerden of the Fisheries Library at Food and Agriculture Organization in Rome; and both Patricia Cook and Craig Wilson of the National Marine Fisheries Service Northwest Fisheries Science Center Library in Seattle. In addition, I would like to thank the librarians who made available the archived resources of the Marine Biological Station at Millport, Isle of Cumbrae, Scotland; the Ministry of Agriculture and Fisheries Laboratory at Lowestoft and the Plymouth Laboratory in England; the Marine Laboratory and the Woods Hole Oceanographic Institution, Woods Hole, Massachusetts, United States; the Institut Français pour l'Exploitation de la Mer at Brest, France; the Cape Piscatorial Society, Cape Town, South Africa; and, finally, the curator of the Preston and Blanche Leigh Collections at the Brotherton Library of the University of Leeds, England.

Next, I would like to acknowledge the help of many friends and colleagues who kindly provided me with photographs and drawings from the archives of their institutions or their personal effects to illustrate the book. These include Rudolph Berka, Czechoslovakia, for permission to reproduce photographs from his collection of early fish farming in Bohemia; Aliaky Nagasawa of the National Salmon Service, Hokkaido, Japan, for pictures of early farming activities in Japan and Chile; the Sumiyoshi Fisheries Corporation in Ino City, Kumamoto, Japan, for its photographs of the Kathleen Drew Festival; Clark Erickson of the Department of Anthropology, University of Pennsylvania, United States, for his aerial photograph of medieval fishponds and fish weirs in Bolivia; and Arnt Ove Olaisen, for his photographs of Norwegian aquaculture.

In addition, I would like to thank the Museum of London Archaeological Services, England, for its permission to reproduce the photographs of the medieval fishponds at Pykes Garden in London, and to the former British Nuclear Fuels Limited Magnox Electric PLC for permission to reproduce the photograph of Hunterston Nuclear Generating Station A in Scotland. I would also like to thank Dale Brown of Seattle for advising on the layout of the maps, and to Nystrom Division of Herff Jones, Inc. for permission to reproduce and use the basic maps from its collection.

I also acknowledge the help of many friends and former colleagues from Czechoslovakia, France, Indonesia, Italy, Norway, Sweden, Taiwan, the United Kingdom, and the United States with whom I checked specific details about some of the more recent aquaculture events. Thanks, also, to Xiao Lu and Mrs. Roland-de Pol, both of the Food and Agriculture Organization of the United Nations, who provided translations of documents in Chinese and Dutch, respectively. In addition, I would like to thank Peter Becker, Bill Fairgrieve, Jeanne McKnight, and Karl Shearer for researching many parts of the manuscript; and last but not least, for the help from my elder son Alistair and his wife Ani for their assistance with the figures and maps, from my younger son Simon for obtaining photographs and researching historical information in China and other Asian countries where he has been teaching business, and from my daughter Emma and her husband Todd, for continuously searching for information in the local libraries.

Finally, I would like to thank Susan Thomas for her editing and correction of my American English, and her professorial preparation of the book. Without her help, it would not have been achieved.

Colin E. Nash
Bainbridge Island
January 2010

Chapter 1

Fish and Shellfish as Food

Abstract

From the remains of Stone Age kitchen middens, archeologists have been able to identify the species that were most commonly eaten by early man. Humans still have a great affinity for seafood and prefer most of the same species. By the Bronze Age, people had developed skills for preserving seafood for transport over considerable distances, allowing trade of such products. Fish were preserved by cooking, drying, and later by salting. When ancient civilizations around the Mediterranean and Asia Minor expanded their trade by ship, salts and spices for preserving food became important global commodities—as they remain today. Seafood is a highly nutritious diet, offering a rich source of protein, as well as unsaturated fats, vitamins, minerals, and trace elements. Although some wild-caught species or their body organs can be naturally toxic, even lethal, today's consumer is in no danger of ingesting such toxins from farm-raised fish.

1.1 Secrets of the kitchen middens

A Neanderthal man stepping into a twenty-first century kitchen at dinner time would easily recognize the aroma of grilled fish, broiled mussels and oysters, or smoked eel as an inviting meal was prepared. There is little or no difference between the favorite seafoods of early man and those that are preferred in most modern households today. From deep in the well-preserved remains of Stone Age kitchen middens, archeologists have been able to identify the species that were the most commonly eaten at that time. In Europe, salmon, tuna, eels, and sea bass were the most typical marine or migratory species eaten by paleolithic

The History of Aquaculture. By C. E. Nash. Published 2011 by Blackwell Publishing Ltd.

humans, and trout and carps were the most popular true freshwater fish in their diet.

Further, paleontologists and marine zoologists, from their studies of fossils in marine deposits, have concluded that in general, the evolution of fish and shellfish has hardly changed these organisms over the last million or more years. This is evident from the drawings on the walls of the caves inhabited by Stone Age man and by the physical features of skeletons of fish or shells of mollusks unearthed from the large volumes of material found in the kitchen middens.

As hunters of common property resources, early tribal societies found many aquatic animals to be very accessible sources of food. Shellfish, such as oysters, mussels, and clams, were available for harvesting at low tides, and fish could be trapped, speared, and netted—all of which was probably quicker and more rewarding than catching either large wild animals or birds. Seafood would also have been easier to handle and cook than were red meats, and not quite as tough to eat.

Archeological traces of primitive cultures reveal a large number of early settlements close to the shore of the sea. Beside ancient menhirs, the large upright stone monuments erected as monoliths or groupings during the Stone Age, archeologists have found many bones of fish and shells of mollusks buried in prehistoric kitchen middens, providing firm evidence that seafood was already at that time a regular part of the diet. The same kind of identifiable remains are to be found in middens beneath the ruins of Greek and Roman occupation throughout their vast empires and under the monasteries and great churches founded in the Middle Ages. The evidence leaves no doubt that man has continued to have a great affinity for seafood for many thousands of years.

One puzzling aspect of archeological discoveries in the kitchen middens of later settlements was that the bones of marine fish were much more in evidence than those of freshwater fish, even though many of these ancient settlements were far from the coast. It would seem that freshwater fish would have been more readily available and easily trapped in inland waters. Nevertheless, perhaps these early settlers found that marine fish and shellfish naturally kept better than did their freshwater catch, or perhaps they had developed skills that allowed them to preserve fish and shellfish sufficiently for transport over considerable distances.

Such skills could have contributed to the existence of trade in marine fish and shellfish that is known to have been active from the end of the Bronze Age onward. As a result of this trade, exotic seafood was popular in Greek and Roman kitchens. Judging by the fish depicted in their mosaics and in the writings of their scholars, one of the most highly prized fish of the epicureans of that era was the herbivorous parrot wrasse. This fish was rare in the western Mediterranean, but it was consequently captured in large quantities by Roman seafarers around the Greek Islands. The species was not only brought back as a catch to be consumed directly, but also, living specimens of the wrasse were shipped home and then released along the coast of Italy. Even before the customs and social habits of the first great civilizations that influenced so much of the world were recorded, some of the more basic skills of preserving and storing

Figure 1.1 Italy, first century; marine fauna mosaic from Pompei, from the *Casa a cinque piani* (House of the Five Floors): common marine life of a rocky Mediterranean coast is portrayed in such detail that the individuals can be identified to genus and species (see Appendix, Table A4). (Courtesy of Classicalmosaics.com.)

food were discovered by early nomadic societies, as they pursued their relentless search for food.

1.2 Processing fish and shellfish

Fish and shellfish offered early human societies a rich source of animal protein to balance diets that were typically high in root and cereal foods. However, it would have been obvious to all early coastal dwellers that any seafood was an extremely perishable food commodity. If a fish was not eaten soon after it was caught, the flesh was quickly covered on the surface with colorful growths of molds and yeasts, which would have produced a noxious, rancid smell as chemical and enzymatic actions broke down the oils and fats, turning the flesh to a watery pulp. In turn, the odor would have attracted flying insects and scavengers. If these early hunting societies were to survive, they had to be able to carry food along with them and to safely store it to sustain themselves in times when conditions were poor and hunting was not possible. Consequently, there was a burden on them to discover measures that would at least slow down the processes of deterioration in their foods.

Preservation of any type of food requires some form of processing or curing to arrest microbiological or biochemical actions that accelerate decomposition.

Curing can also change certain properties of the tissues or flesh without necessarily causing the loss of the food's natural palatability. In some cases, curing can enhance palatability in interesting ways that could add an individual appeal for discerning consumers.

The most basic form of curing is cooking, or heating food directly over an open fire. Cooking foods in this simple way began in the Stone Age, when shellfish were scavenged from the beaches at low tide, and fish were harpooned with wooden spears tipped with arrowheads made of flint. The high temperatures associated with cooking kill the bacteria and thereby retard the processes of decay for one or two days. With the coming of the Bronze Age and tough metal pots, the same effect was achieved by boiling food in water.

A more subtle form of curing is drying, which reduces microbial decay and the buildup of molds. Fish can be dried in the sun or in the wind by simply opening them and leaving them exposed for a time, either on the ground or spread out on crude trestles. Neolithic fishermen, who were the first to use boats and who made fish hooks from animal bones and the parts of large insects, also were culturally advanced in their storage of dried foods. Such basic fish-drying practices are all still evident in primitive jungle societies surviving in remote parts of Southern Asia and South America to this day.

Drying can also be achieved artificially by using heat from a fire. Experimenting with this practice could have introduced primitive fish-eaters to the delights of smoking. In addition to destroying bacteria and proteolytic enzymes while the flesh is being partially cooked, smoking induces chemical processes that can introduce a new taste, flavor, or aroma to the flesh, usually characteristic of the different woods that are chosen as fuels.

The last and most important method of curing is dry salting, which first began to appear among Bronze Age cultures. By this period in history, fishermen of Mesopotamia and Egypt were using woven nets and fishing lines with metal hooks. Dry salting is a simple and effective technique, and it is one that can be applied in diverse ways to suppress microbial growth. Salts in crystal form rubbed into the flesh before it is dried can preserve products for many weeks. If salts are applied instead in solutions of varying concentrations, they can pickle or ferment whole products that can be kept in pots for many months to years. Salting became very important to the early Sumerians and Egyptians. In ancient Egypt, picklers of birds and fish were artisans attached to the temples, no doubt trained by the priests responsible for the mummification of the country's royalty. The Romans produced garum, a concentrated fish sauce, through a process of fermenting salted, pickled fish scraps and small whole fish. It was a very popular condiment, judging by the numbers of garum-filled amphorae that have been identified in old Roman kitchens, alongside those containing wine and olive oil.

Although these are the traditional and basic methods of preservation or processing, there are many and often subtle differences in their use by the different human societies that followed the Stone Age. For one reason or another, some of these differences would have been simply due to choice—the natural human preference for things that had good taste, texture, smell, and color of the flesh. Other differences would have been due to the geographic location of the society, particularly because of the different climates, and consequently, the many possible types of habitation, all of which would have affected and limited the type of protection and storage used for fish.

Other differences would have been based on the local availability of the diverse raw materials for processing. Each locality offered a particular set of fish and shellfish, distinctive types of wood to fuel the fires, and a characteristic composition of salt or other spices that might have been available. Then, the fish and shellfish gathered would have varied by season: that is, at different times of the year, there could have been obvious differences in the quality, texture, color, and other characteristics of their flesh. As a result, early societies would have discovered that in general, each type of fish and shellfish had to be handled and cured quite differently. The most apparent differences were to be found in the curing of nearshore fatty fish, such as mackerels and herring. Much stronger methods of smoking and salting had to be applied to prevent the highly unsaturated fish oils from becoming oxidized and rancid. All the less oily fish and shellfish needed only milder cures.

In time, the choice of food and the preferred processing and cooking practices would have become the accepted way for all the families of a tribe. The characteristic habits of their society would have been subsequently ingrained into their traditions and customs. Finally, a few of them would have become a part of their folklore.

1.3   The importance of salt

The extent to which all these different options for preserving fish and shellfish were used depended as they do today as much on the local demand and preference as on the resources to hand. By and large, primitive coastal communities and island societies with ready access to fresh fish and shellfish all year round had little need to cure a lot of fish for their own use. That which was preserved was sun-dried, wind-dried, or smoked. Inland communities living in temperate regions without the luxury of year-round fresh fish and sunshine relied more on smoking foods for preservation. Those living close to natural solar salt pans, which were found both on the coast and in desert areas, likely would have exploited their opportunity to use salt to preserve fish.

The value of salting became most apparent at the beginning of the Iron Age. With the increasing use and size of ships, the ancient civilizations, especially those around the Mediterranean and Asia Minor, had discovered trade. With the coming of trade among societies, salts and spices for preserving food became important global commodities that enabled most fishing communities to cure and store whatever seafood was important to them, and then to trade any surplus of dried and salted fish, including salted and pickled fish in jars.

Both the Mediterranean region and Asia Minor were rich in natural salt pans and salt mines. The value of owning sources of salt was recognized by the early civilizations, so much so that the great Ptolomy family of Egypt grabbed a monopoly for itself, which helped it to become even more powerful.

Subsequently, most of the early Egyptian sites in and around the Mediterranean came under the control of the Greeks.

For the early Greeks, fishing was not a particularly popular occupation, but as the civilization expanded its influence throughout the Mediterranean and the Black Sea, fishing and eating fish became an important part of the social lives of upper classes. Many small communities around the coast of the two seas at the center of the world's civilization thrived on the catching and trading of dried fish and salted fish with Greece. The Greek armies were known to carry salted fish in pots when they were on the march.

The Romans picked up where the Greeks left off, making fishing and salt processing important industries, which made seafood more readily available to the general public. Fishing was also becoming more skilled, and in addition to the common Mediterranean coastal fish, such as mullet, bream, and eel, the Roman kitchens were used to cooking salted tuna fish, mackerel, conger eel, amberjack, and even swordfish. Few of the ancient recipes used freshwater fish, although it is known that fishermen of the lakes and rivers to the north of Italy sent salted fish to Rome.

By the Middle Ages, both salt and seafood were important commodities that were traded widely. This continued for another five hundred years, until the railways of the Industrial Revolution made possible the overnight deliveries of fresh seafood to the capital cities of Europe. But also during the Middle Ages, freshwater fish became a regular part of the diet of the feudal lords and the more educated people. For the first time, bones of freshwater fish were evident in kitchen middens of the large estates and religious monasteries. Fresh fish played an increasingly important role in the religious life of the monks as it became more readily available, even in locations far removed from the coast. As the monks harnessed the energy of rivers and streams to power their flour mills, they discovered that the storage dams constructed to regulate the flow to the mill were useful ponds for holding trout. Subsequently, they were used to hold a new fish that the monks themselves were spreading around Europe. It was called the carp. The carp would provide them with a regular source of fresh fish, and "stew ponds" soon became a necessary adjunct to the kitchen gardens of all the new monasteries and abbeys that followed.

1.4   Seafood and its nutritional value

For the early societies who hunted for survival up and down the coast, fresh fish and shellfish must have tasted just as good as it does today. But, in addition to the good taste, and unbeknown to these early consumers, of course, seafood was a highly nutritious diet. The large resources of protein in the flesh of fish and shellfish contain many readily available amino acids, such as lysine, methionine, and tryptophan, in quantities comparable with those in eggs, meat, and milk. With their unsaturated fats, vitamins, minerals, and trace elements, all equally important to the human diet, fish and shellfish are considered to be almost as beneficial to the body as mother's milk.

Clearly, quantitative differences in nutritional composition occur among the many large groups of species, and also within groups at different times of the year. In general, freshwater and brackish-water species contain about 14% to 25% protein, whereas marine species contain 9% to 26%. Freshwater species usually have a low percentage of fat (the leanest fish have less than 2.5%), whereas some marine fish may be as high as 20% fat. Compared with animal fats, fish oils contain more polyunsaturated components and are therefore beneficial in reducing the buildup of cholesterol in blood.

Fish and shellfish are both good sources of calcium and phosphorus, but more so when small fish and crustaceans, such as soft shell crabs and shrimps, are consumed whole with bones or shell. Iron and traces of copper are also useful contributors to the general composition, as are the B-vitamins in high proportion.

1.5   Dangers of the diet

The modern but traditional preservation methods of indigenous cultures around the world today are probably indicative of great trial and error that went on in primitive societies as they perfected their processes and built up their experience in storing food. At the same time, they would have learned the hard way that some species of fish and shellfish or their body organs were highly toxic and often lethal. What no doubt was very puzzling to them was that exactly the same types of seafood were safe to eat one day and made everyone ill the next. Furthermore, they found that risks were greater at different times of the year. After harvesting and eating oysters or mussels from their usual beds on a fine midsummer day, everyone sharing the meal might have become drowsy and feverish, and their mouths could have suddenly grown numb. This mystifying experience would have occurred more frequently in summer; hot weather increases the buildup of phytoplankton capable of concentrating dangerous toxins or bacteria to levels that are harmful and sometimes poisonous.

Communities living around the Indian and Pacific Oceans would have come to know, too, that some fish were always extremely dangerous or deadly to eat. Puffer fish and trigger fish had to be carefully avoided, although in Japan the former is now recognized as a delicacy, once the viscera have been carefully removed completely intact by a skilled and licensed handler. Some common reef fish and their predators that served as part of the early islanders' diet would have been more of a challenge, because on rare occasion, individuals of species that were normally fine to eat could become deadly poisonous, when they happened to have been carrying ciguatoxin. This toxin is accumulated in the fish, moving up the food chain as herbivores that are affected are eaten in turn by carnivores, which are eaten by other carnivores, and so on. The toxin is not removed from the flesh by conventional cooking. For some early island societies that depended on seafood, the chances of eating an affected fish would have been high. In modern times, for example, the small populations inhabiting the Maldive and Kiribati Islands have the largest per capita consumption of seafood in the world and could therefore be more likely than others in the tropical or subtropical setting to encounter a ciguatoxic reef fish, however rare the occurrence may be in nature.

Today, the public consumer is in no danger of these or other toxins from farm-raised fish. Fish nutrition has been studied more in depth than has any other fisheries field, and manufacturing artificial fish feeds to very strict formulas is a major component of the animal feed industry. The industry is also strictly regulated, and all feed-producing countries are signatories of the United Nations Codex Alimentarius.

Most captive aquatic animals are almost entirely dependent on artificially prepared diets, together with natural foods that might be available in their enclosures. Artificial diets are invariably high in animal protein (15% to 40%, depending on the age and specific needs of the population), complemented by cereal proteins and carbohydrates, oils, and additives of minerals and vitamins. In recent years, efforts have been made to formulate diets using only polyunsaturated fats and to include approved chemical attractants and growth promoters.

Farmed fish, particularly those grown in accelerating temperature regimes, can be fatty. Unsightly excess fat is sometimes removed by greatly reducing the diet before harvest and by increasing water circulation to increase the fish's energy use.

Off-flavors, mainly caused by the direct absorption of a compound called geosmin from the water, can affect the taste of farmed products. Geosmin is produced by certain species of actinomycetes and cyanobacteria that can bloom under particular chemical and physical conditions. These earthy-muddy flavors are common in catfish and other freshwater pond fish and are now reported in some marine shrimps. These can be easily purged by maintaining stocks in high-quality geosmin-free water conditions before harvesting.

The color of the flesh of farmed aquatic animals is readily changed by additives in the diet. The salmonids, for example, are often fed carotenoid pigments in the diet for some weeks prior to harvest to redden the flesh for increased marketing appeal.

In addition to all the nutritional advantages inherent in fish and shellfish, there are other particular benefits of seafood raised on farms. The nutritional qualities of most species are preserved, because fish and shellfish from farms are almost entirely sold on the fresh fish markets, either chilled or iced. This is not true of the natural harvests. Only about one-third of the natural harvests are sold as fresh products; most are frozen into blocks, preserved in some way in cans or bottles, or reduced into commercial fish meals and oils. Another advantage is that a population of farmed fish or shellfish is invariably uniform in size, or similarly sized individuals can be conveniently harvested as needed. Uniformity is a characteristic greatly appreciated by those who first handle the products for the marketplace, because there is no sorting required for the harvest and shipping of the fish and shellfish to the processors, and standard boxes or containers can be used for more efficient transportation. At the receiving end, those who process the products can automate many of the necessary steps, such as gutting and filleting, by using machines instead of by employing more costly hand labor.

1.6   Into history

So we must ask ourselves, where and when did the movement away from the unstable supply of strictly wild-capture fish begin? When did it change to live storing, to feeding, and then finally, to farming fish and shellfish? Is it a modern phenomenon promoted by the nineteenth century Manifest Destiny movement urging mankind to overcome nature, following Biblical tenets to "be fruitful and multiply?" Did it start in the medieval period, or even earlier in the classical periods when Roman Empires provided the wealth and leisure to experiment with raising live fish and shellfish for food? Or do the roots lie even deeper in early societies, with both the needs and desires of early civilizations to obtain fresh seafood and to transcend the unsustainable hand-to-mouth nature of the hunter/gatherers in developing civilizations? The answer actually lies in all of the above hypotheses, as we shall see.

Chapter 2

Seeds in Antiquity (2000 BC to AD 500)

Abstract

The origin of raising fish and cultivating aquatic plants is credited to early Chinese societies that flourished well before 1000 BC. The Chinese literature described "aquahusbandry" in 475 BC, and common carp culture for food flourished in that form for the next thousand years. Today, China produces more freshwater fish in farms than does the rest of the world combined. China spread simple fish farming throughout Asia; in parallel, other ancient Great Cultures developed methods for holding aquatic animals in captivity. Assyrians engineered river dams, creating lakes that held fish resources, and built fishponds for sacred and commercial use. Fishponds around Sumerian temples dated to 2500 BC. Wealthy Greeks and Romans feasted on marine fish and shellfish, and built complexes of fishponds, vivariae piscinae, *to hold but not to culture such delicacies. These water features eventually became showpieces of great villas. The Romans likely spread similar practices throughout their empire.*

2.1 Ancient origins of fish culture in Asia

The origin of raising fish and cultivating aquatic plants, like that of many other modern technologies, is credited to early Chinese societies that flourished well before 1000 BC. In its rudest form, some systematic sowing and harvesting of fish in China is inferred in the marks on ancient "oracle bones" that have survived. After being burned in fire, these bones, which were frequently the scapulae of oxen or shells of turtles, developed crenations along the edges, which were

The History of Aquaculture. By C. E. Nash. Published 2011 by Blackwell Publishing Ltd.

used by the priests and elders as divinations of future events. Any simple shapes construed as fish were perceived to predict favorable times to gather fish "seeds," and then to replant them in the ponds and lakes of floodplains nearer to home. The seeds were predominantly the young of the common carp, which abound high in the great rivers of China when the rainy seasons begin.

Although oracle bones can reveal little more about these ancient practices of gathering seeds and raising fish, their existence was confirmed by the writing of Si Wen Ming (also known as *Da Yu*, or Yu the Great, founder and first emperor of the Xia Dynasty in China), who reportedly lived around 2070 BC. He also wrote about the laws that regulated the periods during which fish spawn could be harvested. To find government regulation even at that early date is not surprising, because for centuries, fish seeds were an important item of commercial trade, and actual records of fry collection and transportation are found in the writings of Chow Mit in 1243 and again in those of Hsu Kwang in 1639. A history of the Chinese Empire written by a Jesuit missionary in 1735 described the traditional gathering of busy merchants every May on the banks of the Yangtse River to purchase fish spawn. Using mats and hurdles woven from reeds, the peasants covered great lengths of the river to catch the spawn and sell it to the merchants, who then traded it throughout the empire. The live fish markets continue to operate to the present day, but now buyers from the private fish farm communes have replaced the private merchants.

The earliest detailed records of keeping fish in captivity are found in the classic writings of the Chou Dynasty (1112 BC–221 BC), but the actual purpose was not fully explained. The ancient Chinese were never renowned as fishermen with rod and line, or with woven nets, but their early writings indicated many skills with a variety of traps made from the versatile bamboo and marsh reeds. The origins of domestication of fish may have been no more than a practical expedient of having fresh food ready on hand each day, but in early Chinese culture, the common carp was rapidly becoming a symbol of great distinction and standing in the social hierarchy. Therefore, domestication could also have been helped by wealthy landowners and merchants who built ornamental fishponds to beautify their gardens both for their own artistic pleasure as well as for their aggrandizement.

In all probability, all three reasons for the more fortunate citizens of ancient China to keep captive fish on their properties—for food, for ornament, and for status—were well justified. The uses are so closely linked that it would be difficult to distinguish with any conviction which one might have come first, and all are extremely plausible speculations for that time. The common carp has long been a token of good fortune in China, and therefore, a highly acceptable gift between would-be friends. The offering of a live carp or of an image in jade or ivory of the fish rising out of the water to reach the gates of the dragon recognized the increasingly high position and importance of the recipients and bestowed on them great bounty. It is not hard to imagine that both givers and receivers needed a place to keep any such live gifts. Ancient hand-painted scrolls frequently illustrated scenes from domestic life with halcyon gardens complete with ornate ponds full of fish swimming among exotic plants.

The earliest reference in Chinese literature to "aquahusbandry" as a primitive technology is found in the writings of Fan Li in 475 BC. Fan Li was a wealthy merchant and was once the chief minister of the state of Qi before resigning and giving away all his properties to the poor. He moved to Tao, Shandong Province, and changed his name to what is simply translated as the Reverend Mr. Zhu. Reportedly, he made another fortune and earned for himself a reputation for wisdom and generosity. In his short but famous *Yang yü ching*, or *Treatise on Fish Culture*, he wrote of the merits of culturing carp as one of five ways to make a good living in China. He described techniques for constructing ponds just over an acre in size, with "nine islets and eight pits," for selecting adults for breeding, and for breeding, feeding, and maintaining a healthy population. Some of his recommended practices and his ideas of intensive production with regulated harvesting are closely comparable to modern methods of carp culture, and they indicate his keen eye for fish behavior and an understanding of rudimentary biology. Some of his other ideas about life, however, remain clearly far-fetched. It is believed that much of his own wealth, which he seemed to amass easily wherever he lived, was associated closely with the culture of common carp.

On the basis of Fan Li's guidelines, simple common carp culture for food production flourished in China for the next thousand years. Shi Ma Tsen, between 140 BC and 88 BC, recommended the practice to Emperor Wu Ti, noting that a thousand *shik* of fish could be harvested from one pond after only a year of cultivation. Consequently, because all waters in ancient China were publicly owned, fish culture became an integral part of rural life.

Then quite fortuitously, an event occurred that revolutionized fish culture in the country to the present day. In about AD 618, during the Tang Dynasty, an emperor came to the throne whose family name was Li. In the Chinese language, the name for the common carp is pronounced like the word *lee*. Thus, any talk of culturing, killing, and eating *lee* was deemed by Emperor Li's advisers to be grossly insulting to him personally, and therefore, the culture and keeping of common carp by anyone was banned. This crisis, however, forced the peasant farmers to search for other species of fish to culture in their ponds and to develop new husbandry practices. Although there is no record to identify the originators of these new developments, farmers were soon successfully culturing four new species of fish in place of *lee*: the silver carp, the bighead carp, the mud carp, and the grass carp. In addition, the wild goldfish proved to be much more suitable as a progenitor of beauty for the breeding of ornamental fish, a practice that by then had become popular.

The four new types of fish that replaced common carp for food production in the rural areas coincidentally had different dominant behavioral habits with regard to their own nutrition. Silver carp primarily fed on zooplankton and phytoplankton; the bighead consumed phytoplankton and decaying vegetation; the mud carp fed on detritus; and the grass carp grazed only on macrovegetation. Although there was some overlap in diet, each of these species occupied a different niche in the aquatic environment without too much interference from the other three species. As a result, the four types of fish could be cultured together in the same pond.

With the interest in these new fish and their different feeding habits, farmers soon understood the importance of maintaining high productivity of the water and enriching it with animal manure and organic wastes. Consequently, fish culture in China became more closely tied to animal and plant husbandry, and ponds were built to receive manure from pigs and poultry that were housed in buildings around or directly over the water. Furthermore, the banks of the ponds also proved to be fertile and well-irrigated places for growing plants and vegetables. In particular, the farmers planted rows of young mulberry trees, the leaves of which were needed to feed silkworms that were also being farmed intensively. Fine silk was much in demand by an expanding textile industry, because it was a valuable commodity for trade. Where fishponds could be drained, the enriched sludge was excavated and used as fertilizer for crops. Not surprisingly, with such close attention to nutrient sources and pond productivity, the yields of fish increased dramatically, and polyculture of carps in these integrated farming systems became a cornerstone of rural life in China. It remains so to the present day.

It is surprising that the existence of fish farming in these areas was not mentioned by Marco Polo in his travelogue through China and Southeast Asia in the thirteenth century. He described in great detail the many kinds of animals, birds, trees, and vegetation that he saw on his phenomenal journey, and because of his writings, it is often inferred that he was a great lover of nature. Yet, Marco Polo never specifically described fish. He related his frequent confrontations with many large rivers and lakes, and remarked on the great systems of canals that were used for shipping and defense, and those that "carry off filth to the sea." He marveled that fish were shipped daily from the seacoast and far up the rivers to the local markets in a quantity so vast that it would seem impossible to sell them all, but he never identified any of the fish, even crudely. He commented only that the varieties changed according to the season of the year.

Later, while en route around the coast of India, Marco Polo described the traveling merchants who hired pearl fishermen to work the oyster beds, but nowhere on his journey through Asia did he mention fish that were raised in ponds. Perhaps as he moved down through China, he did not realize that many of the fish he saw and ate were being farmed and that he was missing the opportunity to describe yet another practice among many that originated in that part of the world. Today, China produces more freshwater fish in farms than does the rest of the world put together, and it is almost entirely through polyculture of *lee* and the same four cyprinid fish that were raised together in past millennia.

2.2   The *vivariae piscinae* of the Great Cultures

The spread of simple fish farming throughout Asia in the last two thousand years is also credited to the Chinese, although in all probability, there was some parallel evolution through extension in all of the Great Cultures of the time. The subcontinent of India was invaded countless times from the north, and its early civilizations were crowded along the two great valleys of the Ganges River and the Indus River. The early occupants of these fertile valleys were infatuated with water and fastidious about cleanliness, and they developed quite elaborate sanitation systems, complete with reservoirs, holding tanks, and piping. They also used the reservoirs to maintain stocks of fish. The Indian philosopher, Kautilya, who lived in about 300 BC, mentioned the keeping of fish in reservoirs and how to poison them should they be taken in the time of war. Kautilya was not unlike Fan Li in his respect for material success. In his philosophy, described in the relatively recently discovered *Arthásastra*, he expounded four ways for personal prosperity through the use of wealth and the land.

Further references in Indian and Chinese literature do not appear again until much later. King Somesvara, in his great tome, *Manasollasa*, written in 1127, included a chapter on fishing and methods for fattening fish in ponds. This was followed in 1243, in the Sung Dynasty, by Chow Mit's description of methods for the transportation of young fry in his book *Kwei Sin Chak Shik*.

The purpose of simple fish husbandry in the Great Cultures could have been mainly the expediency of keeping fish and shellfish alive and fresh for the table, or for some religious function and symbolism. There is no evidence that fish were continuously bred in captivity, and little indication that they were owned and raised in ponds by the common people. In Egypt, there is a bas-relief on the rock tomb of Akihetep from about 2500 BC, which shows fishermen catching tilapia, catfish, and other fish of the Nile River with a net complete with floating head-line. Because of the square edge to the water below the fishermen, it has been suggested that the fish are in a pond, and not in the open river—which may or may not be the case.

Although the Egyptians are recognized as the first to develop angling for fish, they were not renowned as fishermen. Mostly they confined their efforts to catching the abundant resources of the Nile and its great delta. However, they did build *vivariae* adjoining their temples and palaces to house living creatures, and particularly some large *vivariae piscinae*, or pools, for holding aquatic animals, such as reptiles and amphibians as well as fish. The behavior of these sacred animals kept in captivity in the temples was interpreted as a serious omen for the time, and individual creatures, such as a large eel, terrapin, or crocodile, were frequently decorated with gold and precious stones to encourage good portents.

Ancient Egyptian writings also indicated that the people understood the principles of migration and breeding of fish in the Nile, but there is no evidence of any attempted domestication of the fish in captivity. Although fish were an important staple in the commoners' diet, they were eschewed by Egyptian kings and priests for symbolic reasons. Some types of fish were sacred, because they were believed to guide the boats that bore the dead to eternity; others were avoided, because they were regarded as unclean. The latter were probably the muddy-tasting freshwater species from the slow-moving waters in the outer reaches of the Nile Delta.

The Assyrians of Western Asia, including both the Sumerians and Babylonians, were greatly fond of fish and were more skilled at fishing. In their apparent preoccupation with water and hydraulic engineering well before 2000 BC, the ancient Semitics were known to keep large resources of fish in lakes created by the construction of dams on the Tigris and Euphrates rivers to improve the irrigation of their crops. They, too, built sacred *vivariae piscinae* for their temples, but also *piscinae*, or simple fishponds, to provide food and to make a commercial profit. Virtually every Assyrian township had its own fishponds. Records of 422 BC described fishponds belonging to rich merchants and civic prefects that operated under contract for one-half *talent* of silver and a daily supply of fish for their tables.

Ponds for both sacred and commercial fish around the Sumerian temples can be traced back to 2500 BC. Each pond had a keeper who taxed the public to fish in the commercial pool. Further evidence indicated that small ponds became quite numerous among the Assyrian commoners, but not all were protected by the earliest known regulations regarding fishing rights and ownership. It was noted, for example, that fishponds built by "poor men" were often poached, and that there was no legal redress.

The neighboring Israelites, in spite of their propensity for fish and links with Egyptians and Assyrians, both hostile and friendly, never constructed *vivariae* or fishponds. There is no record of *vivariae* or *vivariae piscinae* in all their prolific writings until almost the beginning of Christianity, when they were probably influenced by the practices of the Mediterranean civilizations.

Wealthy Greeks and Romans, who were well known for their epicurean pursuits of the exotic, feasted on the entire range of marine fish and shellfish common to the Mediterranean Sea, for which, at times, they paid extraordinarily high prices. Because of the geography of Greece and the great lack of fresh water in the country, the Greek diet emphasized the local marine fish that they could catch conveniently close to their rocky coastline, or obtain, salted, from their colonies around the Black Sea and North Africa, and through commercial trade. Indeed, the ancient Greeks were extremely fond of fish, both swimming in the sea and prepared on the table, and could list some four hundred different types. Nonetheless, there is no record of their building *vivariae* in their temples to appease the deities, or *piscinae* around their great houses to supplement their diet.

In contrast to Greece, Italy was blessed with a long flat coastline, greatly interspersed with vast lagoons and estuaries, and a hinterland with rivers and lakes filled by myriad springs. Consequently, both the Etruscans and later the Romans, both rich and not so rich, were very familiar with the delights of fish and shellfish that thrived in all their waters. The wealthiest citizens who had villas on the Tyrrenean coast both north and south of the capital city built complexes of rectangular tanks along the foreshore capable of holding a variety of future delicacies. Even the wealthy Romans who had country villas far inland in the hills of Umbria to escape the mosquito-infested marshlands along the coast were prepared to collect and transport their favorite seafoods, such as red mullet, oysters, and cockles, over great distances at significant cost so that they would not be deprived of such luxuries for their tables.

Although the *piscinae* of the Romans were costly to build and maintain, as noted by more than one writer of the time, they proved to be both extremely

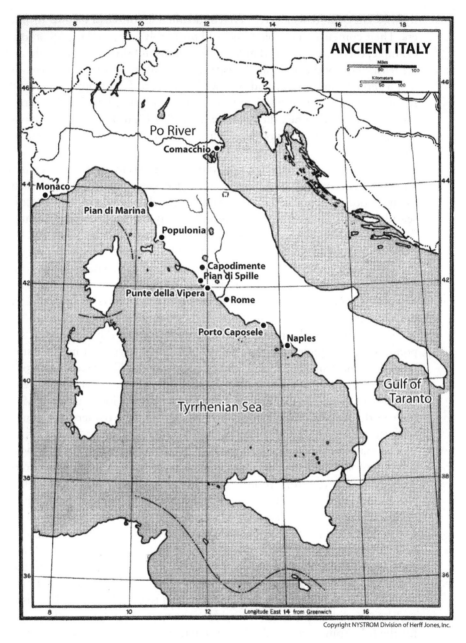

Figure 2.1 Ancient Italy. (Adapted from basic map, copyright NYSTROM Division of Herff Jones, Inc.)

popular and a social phenomenon. Their original purpose, that of keeping seafood fresh in a hot climate, was soon replaced by that of being prestigious showpieces, complete with privileged keepers who hired fishermen to catch sprats to feed the ponds' inhabitants. Therefore, these *vivariae piscinae* remained the prerogative of the wealthy patricians and were privately owned and protected by law. They could also have served as a valuable source of income for the owners. Their contents were able to be sold profitably either in large quantities—for example, there are records that some five thousand thrushes were sold annually for feasts—or by the individual creature, particularly if a large fish, eel, or turtle had a reputation as a pet and was well trained to feed from the hand. But for the most part, *vivariae piscinae* were simply luxuries, which led Cicero to rail bitterly against what he called *piscinarii*—a derisive term for fishpond owners or enthusiasts—and the money they lavished on the construction of their pleasure gardens.

The *piscinarii* developed the Egyptian- and Assyrian-style *vivariae piscinae* to an extraordinary scale and made them a necessary part of most of their great estates, both maritime and inland. Many were elaborate. Plutarch described the seaside villa of L. Licinius Lucullus on the Bay of Naples, where Lucullus had "surrounded his buildings with circuits of the sea and channels for breeding fish." Cicero, who boasted of his many villas, each for a different season of the year, took great delight in their thermal baths, promenades, aviaries, and fishponds with fountains. The archeological complex of Porto Caposele near Formia in southern Lazio shows that the villa had two separate *piscinae* (or *peschiere* in Italian) that seem particularly grandiose and well planned, whereas at Pian di Spille to the north of Rome, the layout of the rectangular seawater ponds appears simple and more purposeful. Both were built in the first century BC. The semicircular *piscinae* at the Etruscan villa at Grottacce dates to the first century AD and adds a decorative touch to the coastal villa, but the extensive complex of ponds and walkways of the Punte della Vipera at Santa Marinella, built in the second century AD, shows clearly that the owner took aquaculture seriously.

The majority of the Roman ponds that were constructed on land were built either outdoors in large open areas or inside large pavilions. They were frequently adjacent to the refectory rooms to enable guests to see and choose their own live animal, bird, or fish to sacrifice and consume. The more ostentatious were designed around the *triclinium*, a place where the patricians lounged to talk and to eat. At the villa of Sperlonga, which was probably built by one of the wealthy *piscinarii* on the coast south of Rome in the first century BC, the triclinium was constructed on a small island in the middle of a large pool in a grotto of the cliffs. The pool acted as the dining table on which food was floated around in small ships kept filled by the servants. Adjacent to the island were four fishponds, each of which had pipes embedded into the walls as a refuge or breeding place for the fish. Some two hundred years later, when the villa was often frequented by Emperor Tiberio, a well-described landslide occurred at the grotto, which almost took his life as he dined in the triclinium. Another more elaborate *piscina* for marine fish was built on board an exotic trading ship that enabled the diners to cruise around the bay. It consisted of a twenty-one thousand-gallon seawater tank kept filled with freshly aerated water, aided by the engineering skills of Archimedes and his famous screw pump.

Freshwater *piscinae* were used for holding, rather than for raising, fish, although there is a suggestion from Sperlonga and other villas that pots sunk into

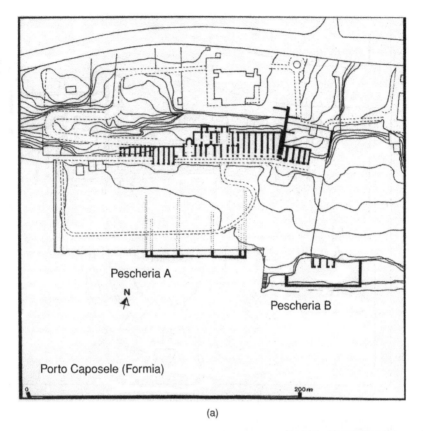

(a)

(b)

Figure 2.2 Italy, first century BC; plan of a Roman villa and *pescherie* (or *peschiere*): (a) two *pescherie* at a Porto Caposele villa, Formia; (b) remains of *pescherie* at Pian di Spille, Tarquinia. (From Schmiedt, G. [1972] *Il livello antico del Mar Tirreno*, Casa Editrice Leo S. Olschki, srl, Florence, Italy. Courtesy of Casa Editrice Leo S. Olschki, srl, Florence, Italy.)

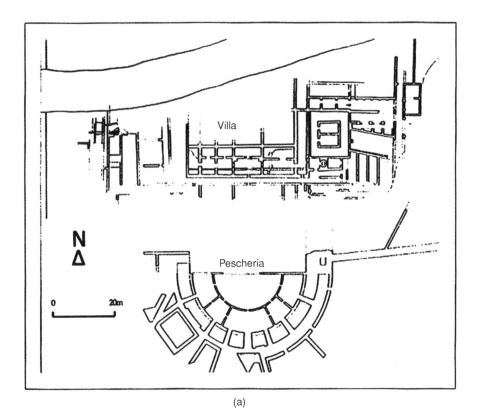

(a)

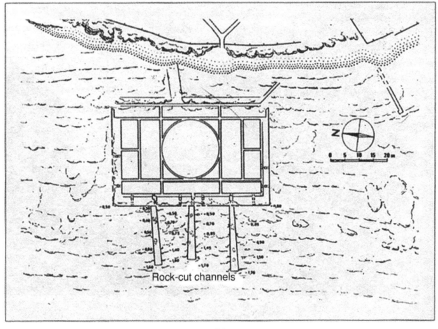

(b)

the walls of the ponds possibly were provided as potential places for the fish to spawn. Availability and cost of freshwater probably limited the use of ponds through the year. Although highly skilled in hydraulic engineering and capable of moving vast quantities of water into the cities and farmlands by a great network of brick-built aqueducts, the Romans still had to pay for their supply. In some respects, water was also rationed; urban aqueducts usually operated only at certain times, and large cisterns had to be filled to bridge the intervening days. Consequently, in the hot, dry climate of the Mediterranean, the high evaporation rates would have made the use of water in open fishponds a significant waste. Only patricians fortunate enough to live in villas near spring-fed lakes were able to build swimming pools, gardens with water fountains, and ponds for showing off their domesticated fish.

Nonetheless, the ordinary Roman people were not deprived of fish. The rich brackish-water lagoons and estuaries that surrounded the coast of Italy were public waters, where the natural shoals of gray mullet, sea bass, and sea bream lingered to feed and were subsequently trapped inside. The common people also harvested the large resources of eels that moved up and down the rivers on their seasonal migrations. Clearly, the coastal waters of Italy were rich, and the observant Pliny noted that waters containing eels were always clean and healthy.

Many Roman *piscinae* were used for shellfish. This is unusual in view of the more perishable nature of such species in poor conditions. But the Romans were particularly partial to oysters, according to Pliny. They collected them from the coastal waters around the geographic heel of Italy and cultivated them artificially in Lago Lucrino, Lago Fusaro, and in the great Gulf of Taranto. Their crude methods of culture have been found depicted on several old Roman vases and etched on glass bottles. They also brought oysters back to Italy from all parts of their empire. The writer Dio Cassius commented that although the northern seas beyond Gaul (France) were swarming with fish, the natives never ate them—a situation of which the Romans readily took advantage, evident from the abundant remains of shellfish that were left in the kitchen middens of most of their sites. The Romans were also known to manage oyster beds in France and England and to carry live oysters and even fertilized eggs back to Rome.

There is evidence that the Romans also spread the practice of keeping fish and shellfish in *piscinae* throughout their empire. The Roman rural villa at Montmaurin, for example, close to the valley of the Garonne in southwestern France, had an extensive garden for the production of wine and other crops, and special ponds for oysters and shellfish. A poet writing in about AD 550

←———————————————————————————————

Figure 2.3 Italy, first century AD; Etruscan and Roman *pescherie* (or *peschiere*). (a) Plan of an Etruscan villa and *pescheria* at Grottacce. (From A. Bufalo, in De Rossi, G.M., Di Domenico, P.G., and Quilici, L. [1968] La Via Aurelia da Roma a Forum Aureli, *Quaderno dell'Istituto di Topografia Antica della Universita di Roma* **4**, Figure 152. Courtesy of Ufficio Stampa e Comunicazione Sapienza, Università di Roma, Italy.) (b) Line drawing of remains of Roman *pescherie* at Punte della Vipera, Santa Marinella. (From Schmiedt, G. [1972] *Il livello antico del Mar Tirreno*, Plate 83, Casa Editrice Leo S. Olschki, srl, Florence. Courtesy of Casa Editrice Leo S. Olschki, srl, Florence, Italy.)

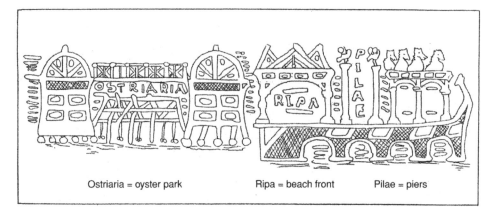

Ostriaria = oyster park   Ripa = beach front   Pilae = piers

Figure 2.4 Italy, late third or early fourth century; portrayal of Roman oyster park (*ostriaria*): line drawing of the Roman Populina Bottle, abraded decoration on transparent, pale green, blown glass (H 18.4 cm). (Courtesy of the Collection of the Corning Museum of Glass, Corning, New York [62.1.31].)

described fishponds in country villas in Burgundy, at that time likely occupied by Romanized barbarians. However, even though many of the foundations of excavated Roman villas and fortifications outside Italy include the remains of cisterns and pits that had been made watertight with plaster, these would have been only for storing drinking and irrigation water. Fish and shellfish would have been excluded. Real fishponds, if they had been built, would have been sited in low marshy ground away from the house, which would have made it highly unlikely for any evidence of them to survive.

More practical *piscinae* in this ancient world were probably the defensive moats of the larger walled fortifications and cities, which like those of ancient China, would have contained fish by accident or by design. The water in these moats was highly enriched with human and kitchen wastes dropped from openings in the walls above, which would have provided good nourishment for certain types of freshwater fish, such as the carps. Yet, on the whole, the famous historians of the times, such as Horace, Pliny, Seneca, and Cicero, waxed eloquently on the delights of marine species: red mullet, the king of fish; wrasse; sole; turbot; and oysters. Freshwater fish were for the common people. Aristotle, who is recognized as one of the earliest naturalists, and Pliny both wrote that the carp was not held in high esteem by the Romans. Consequently, if the keepers of the ancient *vivariae piscinae* had developed any artificial system of breeding and cultivation of fish, some account of it likely would have been found among the works of these prolific writers of the times.

Bibliography

Aristotle (fourth century BC) and Pliny (first century AD) [carp not held in high esteem by Romans]. Cited in: Houghton, W. and Lydon, A.F. (1879) *British Fresh Water Fishes*, vol. 1. W. Mackenzie, London.

Chow Mit (1243) *Kwei Sin Chak Shik*. China [in Chinese].

Cicero, Marcus Tullius (60–59 BC) [railed against piscinarii; described his villas] *Epistulae ad Atticum* (Letters to Atticus), I.20, I.19, II.9. Translated in: (1999) *Cicero: Letters to Atticus*. (translated by Shackleton-Bailey, D.R.). Loeb Classical Library, Harvard University Press, Cambridge, MA. Cited in: Karl-Wilhelm Weeber (2003) *Luxus im alten Rom: Die Schwelgerei, das süsse Gift* (Luxury in ancient Rome: Indulgence, the Sweet Poison), Chapters 3 and 4. Primus, Darmstadt.

De Rossi, G.M., Di Domenico, P.G., and Quilici, L. (1968) La Via Aurelia da Roma a Forum Aureli. *Quaderno dell'Istituto di Topografia Antica della Universita di Roma* 4.

Dio Cassius [northern seas swarming with fish, which natives did not eat]. Cited in: Foster, H.B. (1914) *Dio's Roman History*, 9 vols. (translated by E. Cary). Loeb Classical Library, Harvard University Press, Cambridge, MA.

Horace, Pliny, Seneca, and Cicero [waxed on delights of certain edible fish]. For example in: Pliny (1940) *Natural History III*, Books 8–11. (translated by I.L. Rackham). Loeb Classical Library, Harvard University Press, Cambridge, MA.

Hsu Kwang Chi (1639) *The Complete Book of Aquiculture* (Ming Dynasty) [in Chinese].

Jesuit missionary (1735) *Description géographique, historique, chronologique, politique et physique de l'empire de la Chine et de la Tartarie chinoise* (Geographic, Historic, Chronological, Political, and Physical Description of the Chinese Empire and Chinese Tartary [Northern China, Japan, Korea]). Jean-Baptiste du Halde, Paris [in French].

Kautilya (ca. 300 BC) *Arthashastra*. In: Kautilya's *Arthashastra*, a treatise on ethics and righteousness in foreign policy and intelligence and its relevance in modern times. Paper presented at the annual meeting of the International Studies Association 49th Annual Convention, "Bridging Multiple Divides," Hilton San Francisco, San Francisco, CA, March 26, 2008.

King Somesvara (1127) *Manasollasa*. Cited in: Sadhale, N. and Nene, Y.L. (2005) On Fish in Manasollasa (ca. 1113 AD). *Asian Agri-History* 9 (3), 177–199.

Marco Polo (1293–1299) *Il Milione* (Travels). [Dictated to a cellmate in prison; subsequently translated, embellished, copied by hand, and adapted. There is no authoritative version.]

Pliny (ca. 77) [Romans partial to oysters]. *Naturalis Historia (Natural History)*, vol. 9. Rome [in Latin]. English translation: Pliny: *Natural History*, vol. IX, Books 33–35. Loeb Classical Library, 1952, No. 394, translated by H. Rackham, Harvard University Press, Cambridge, MA.

Pliny (ca. 70) [noted that waters containing eels were clean and healthy]. Cited in: (1852) Eels. *Fraser's Magazine for Town and Country*. 45 (January–June), 645.

Plutarch (first century) [described the seaside villa of L. Licinius Lucullus] *Bioi Paralleloi* (Βίοι παράλληλοι, Parallel Lives), Greece [in Greek]. Also cited in: Higgenbothem, J. (1922) *Piscinae: Artificial Fishponds in Roman Italy*. University of North Carolina Press, Chapel Hill, NC.

Poet (550) [described fishponds in Burgundy]. Cited in: Fussell, G.E. (1969) The classical tradition in West European farming: the sixteenth century. *The Economic History Review, New Series* 22 (3), 538–551.

Schmiedt, G. (1972) *Il livello antico del Mar Tirreno*. Casa Editrice Leo S. Olschki, SRL, Florence [in Italian].

Si Wen Ming (2194 BC) *Shan Hai Jaing (Mountains and Seas)*, vol. 18. Yang City, China [in Chinese].

Chapter 3

Subsistence Farming through the Middle Ages (500–1450)

Abstract

Early in the ninth century, simple fish husbandry practices were extending throughout Asia, the Pacific, and Europe. The "stew ponds" of European monasteries held live fish for fresh food, and beds of shellfish were protected and managed. China exported its knowledge of fish farming (and its carps) along with other treasures and inventions via naval expeditions during the tenth through fifteenth centuries, until the Great Withdrawal of the Chinese fleet in 1433. Influenced by Moslems who dominated India and controlled trade throughout the region about the year 1000, Southeast Asian island communities began to trap marine fish in coastal ponds and transfer them to shallow seawater tambaks for food production; modern versions continue today in Indonesia. On the other side of the world, engineered mud channels and earthen ponds covering hundreds of square miles of Bolivian river basin likely provided infrastructure for ancient floodplain farming methods that could have yielded fish year-round.

3.1 Introduction

The Middle Ages began in about the year 500 with the ending of the Roman Empire in Western Europe and lasted until about the middle of the fifteenth century with the advent of the Renaissance. Until the coronation of Charlemagne in 800 and the revival of the Holy Roman Empire, the early centuries of the Middle Ages were collectively known as the Dark Ages. And with good cause. Urban life broke down with the lawlessness of the warring tribes who moved

The History of Aquaculture. By C. E. Nash. Published 2011 by Blackwell Publishing Ltd.

freely through the shattered remnants of Roman occupation and orderliness in Western and Central Europe, destroying almost everything in their path.

Fortunately, Eastern Europe remained strong with the growth of the Byzantine Empire, allowing monasticism to flourish in the region between the fourth and the eighth centuries. This was the founding period of the Carolingian Renaissance, which was to become the most powerful force in learning in the restructured societies that developed throughout Europe under Charlemagne. One small but vital part of the cultural rebirth was the use of simple aquaculture.

3.2   "Stew ponds" of the great estates in Europe

The common carp was mentioned by name in *Capitulare de Villis*, a book written by King Charlemagne in 812. Hence, at the beginning of the ninth century, the practices of simple fish husbandry were extending throughout Asia and moving out into the Pacific, and concurrently, were conveyed to Europe and put into use. It has been hypothesized that the carp was introduced into Europe from Asia through Cyprus—hence the derivation of its Latin name *Cyprinus carpio*. But this has never been verified. In all probability, the basic Asian techniques for husbanding and then for raising carp were reinforced and improved by some skills independently developed in parallel on the great estates and isolated monasteries that had survived the ravages of the hordes. But with the introduction of carp to Europe, stew ponds became increasingly more common as the Dark Ages slowly brightened. They provided a convenient way to hold a source of fresh protein-rich food to supplement a diet of dried, salted, and pickled foods. All of these were important for the basic survival of small communities, not only through bleak winter months, but throughout those lawless times, frequently beset by siege and war.

Similarly, beds of sessile shellfish, such as oysters and mussels, which had been consumed by man since prehistoric times, were once again protected and crudely managed by some selective harvesting. The *boucholeurs* of France, for example, originated in the thirteenth century. Supposedly, an itinerant Irishman called Walton discovered that the nets he left spread out on the shore for the capture of wading birds were soon covered with small mussels, and these mussels continued to grow once they were protected from predators in the mud. This observation led to the bouchot system of growing mussels on wooden stakes interwoven with twigs to form a net. These were then laid out in neat rows to be alternately covered and exposed by each tide. The tall mussel stakes of the *boucholeurs* were known to cover many acres along the west coast of France. One particularly popular place became known as the Anse de l'Aiguillon, or Bay of the Stick.

As in the earlier times of the Great Cultures centuries before, the fishponds built and used throughout the Middle Ages were all private, and ownership and operation continued to be the privilege of priests and nobles. Stew ponds are evident in the earliest records of many religious orders in Europe, most of which were Roman Catholic. Although fish were largely avoided in the diet of the early monastic orders because they symbolized impurity, the eating of fish suddenly became an important daily part of cloistered life when the Pope in Rome decreed a very large calendar of religious fasts and feasts, all of which were strictly observed by the Church and the people, alike. In England, the monastic calendar featured no less than 145 days of observance. The Lenten festival of St. Ulric, for example, was known for the consumption of carps, pike, and mullets, and the monastery at Grandjilla, near San Lorenzo de El Escorial in Spain, still produces "lean meat" for Lent from its fishponds. Stew ponds for overwintering fish for the kitchens are visible among the ruins and foundations of many of the earliest Catholic monasteries in Europe, after the buildings were destroyed in the Reformation of the sixteenth century.

Among the entries in the *Domesday Book*, the great inquisition of England ordered in 1086 by William the Conqueror to register the extent and details of all the lands and properties of his newly acquired kingdom, the Abbey of St. Edmonds was recorded as having access to *"II vivariae piscinae"* for the provision of the refectory. The *Domesday Book* also made reference to many tidal fisheries of England and to tenants who held weirs by the sea for which they paid an annual rent to the local county borough. These weirs were a source of salmon and eels and probably gray mullet feeding in the rich estuaries. Early records and accounts for the weirs along the River Severn revealed that the tenants were called "farmers" who "farmed the fishery."

The importance of fish in monastic life throughout the Middle Ages is evident from many surviving documents. The original charter of the Kladruby Monastery in Bohemia, for example, which is dated 1115, described a fish production pond for carp in great detail. Later, in 1854, one Baron Mongaudry was hunting in the archives of the French Abbey of Réome, now Moutiers-Saint-Jean, in Bourgogne. There he discovered the old diaries of a monk called Dom Pinchon, who had experimented in 1420 with a hatching box in streams for the incubation of fertilized eggs of trout that he had been collecting.

The Catholic Church kept its finger in the pie of most of the financial enterprises of the Middle Ages to share in the revenue. In Italy, for example, since the days of the Romans, the people living along the northern Adriatic coast had been constructing dikes and gates to defend themselves and their properties, which were scattered along a littoral zone, against the continual threat of flooding. Over time, with each flood-control measure, the inhabitants created huge fenced enclosures, or *valli* (derived from the Latin word for paling, which is *valleum*), filled with labyrinths of lagoons, ponds, and drainage canals. These became characteristic features around most of the great lagoons and river estuaries. The long evolution of the famous *valli* around the mouth of Po River, for example, is well documented in the archives of the town of Comacchio. By the beginning of the thirteenth century, the *valli* were becoming increasingly organized into an extensive cooperative and social system for the fishermen. Under the military-like command of a "farmer general," who, with his brigade of men, was responsible for all the construction and management of the infrastructure and water control, the Comacchio *valli* contained over four hundred public and private fish farms. Each farm was managed by a chief cultivator and a group of laborers. Nonetheless, the cooperative, which had built and managed this unique *valli* system over centuries, paid an annual rental fee to the Pope for the "privilege of this fishery."

Stew ponds were also a common feature of the large country estates, all of which were in the hands of the aristocratic members of the society. Not surprisingly, these estates occupied prime fertile land that was well irrigated by rivers and streams. The eleventh century *Domesday Book* also recorded large numbers of *vivariae piscinae* on smaller estates of ordinary landowners and other wealthy middle-class citizens in England, on which the king assessed a tax of part of the crop—invariably, eels or pike. In the prologue to *The Canterbury Tales*, written between 1387 and 1392, Geoffrey Chaucer described a franklin, or landowner who was not ennobled, who "hadde many a breem [bream] and many a luce [pike] in stuwe [stew pond]."

The feudal system of the Middle Ages, which specifically deprived the peasants of ownership of land, also deprived them of owning stew ponds and even of having access to fish. Most rivers and streams belonged to kings and their loyal barons, who controlled large territories of land as well as all the game and fish in them. Poaching by the peasantry was invariably on pain of death, as was the crime of stealing fish from stew ponds. There were few inland waters where fish were considered to be common property. Fresh fish to eat was therefore a rare commodity for the lower classes and the rural peasants, and not cheap to buy.

The ancient laws of the Middle Ages regarding fish and fishing began to break down with the signing of the *Magna Carta* in 1215. This constitutional charter enforced on the English Crown by the knights prevented any further granting of specific fishing rights on salmon rivers and ordered destruction of many of the weirs built by the landowners to create large fishing pools. Other European countries would follow this lead eventually, but not all at once, and many feudal restrictions would remain on some legal statutes in Europe, such as those in Hungary, until the end of the nineteenth century. Nonetheless, the family stew pond was one of the first steps in clear and private ownership of that which was a common property resource.

Community fishponds, however, were not uncommon in Eastern Europe, and like the municipalities of the Assyrians over a thousand years earlier, many townships in Bohemia began to build large ponds, often in a spirit of civic competition. The movement was begun by the enlightened Charles IV, a Roman emperor and king of Bohemia who reigned in the second half of the fourteenth century. Conscious of the importance of conserving water and food resources, he commanded estate holders and cities to build fishponds "so that the kingdom would abound in fish and mist." These ponds, many of which were over five hundred hectares in size, not only provided fish for the local population, but also were an attractive environment for wild game and waterfowl that were hunted, adding some welcome variety to the usual diet of salted and dried domestic meats.

Toward the end of the Middle Ages, fish typical of the stew ponds in Europe were native freshwater species, such as bream, perch, carp, barbel, roach, dace, and minnows. Most predatory fish, such as pike, tench, eel, and lamprey, were excluded if possible or kept in separate ponds. The flesh of the common tench was in fact thought to be particularly unwholesome, but it was kept available in ponds because of its purported medical properties. Early physicians prescribed the touch of a tench as a cure for the ague, and so it was used like a large plaster on the sole of each foot of the patient to draw out a fever.

Carps were not recorded in stew ponds in England until the fourteenth century. Early preferences were more for the flesh of eel and pike, which was firm and tasty, and accordingly, these fish were very popular to eat. In 1999, some fourteenth century fishponds were discovered during an excavation beside the River Thames in London. The ponds were elaborately constructed with timber uprights and sides made from English oak. Each pond was over 10 meters long, 3 meters wide, and about 1.5 meters deep. The revetment of one pond was made with a side of a thirteenth century rowing galley, which is clearly identifiable by its portholes framed for the oars. Later, probably a century or more, the timber sides were replaced with chalk blocks, which would have been mined locally, and wattle fencing. This specific area beside the river was once known as the Pyke Garden, and its name was probably derived from the fishponds that held pike and other river fish ready for sale in the nearby market. By the fifteenth century, the ponds appeared to have been abandoned. Without use, they silted up and became a repository for rubbish, which later proved to be a significant treasure trove for the Museum of London's archaeologists.

With the exception of any migratory species, such as salmon, many fish would have remained in stew ponds through the winter months. If the ponds had been suitable and large enough, these fish probably would have reproduced in the following spring, thus replenishing the stocks naturally. From there, it would have been a short step to the discovery that separate ponds were useful for storing over winter, breeding in the spring, and fattening in the summer and autumn.

3.3   The *tambaks* of Southeast Asia

The practicalities of that which might now be described as fish farming spread throughout the ancient world most rapidly after the tenth century, although probably it was still little more than competent husbandry of fish in captivity and simple management of sessile shellfish beds. At that time, the great influence of China was being reinforced and strengthened throughout Asia by a sudden twist in the attitude of the Chinese. Rather than continuing the militaristic ways of the past, the leaders of the new and strongly moralistic Song Dynasty (960–1126) opted for peace and orderliness. To do this, they established an extensive civil service. Although it was not popular because of the heavy taxation, it proved to be a foundation for almost five hundred years of enormous development not only in China, but throughout the Orient. As the opportunities for trade expanded, the leaders of each dynasty continued to export some of their wealth and knowledge quite freely to all those who traded with them.

The largesse of each new Chinese leader would have been immeasurable, because the country led the world in developing astonishing technologies, such

(a)

(b)

Figure 3.1 England; medieval fishponds in London and Pyke: (a) woodblock drawing of medieval London and the River Thames with the fishponds in the foreground; (b) excavation of thirteenth century fishpond at Pyke Garden made of oak timbers, wattle fencing, and chalk blocks. (Courtesy of the Museum of London Archaeology.)

as firearms and printing, which the modern ages now take for granted. They organized the mining industry for the production of steel and improved the design and construction of large seagoing vessels together with the means for their navigation. Great naval expeditions sailed throughout every inhabited land bordering the China Sea and Indian Ocean for centuries, until the Great Withdrawal of 1433. One Chinese Muslim military leader, the Grand Eunuch Zheng, made seven great naval expeditions over a period of twenty-five years, with hundreds of ships. The principal ships in his convoys were extraordinarily large, with several decks and rows of pole masts with cross-trees from which hung the traditional sails of a junk. The lower decks were frequently filled with animals and plants, and in some vessels, the holds could be flooded to carry fish.

Zheng and his admirals spread the word of the grandeur of each new emperor of the Ming dynasty all around Southeast Asia and across the Indian Ocean to East Africa. Their missionaries distributed Chinese treasures and taught Chinese skills wherever they landed. Because of their symbolism as a valuable token of respect, carps would certainly have been a part of such tributes, probably together with simple lessons in fish husbandry. Such an activity would have been aided in no small way by the hardiness of most carps for travel, and by their tolerance of very poor water quality conditions, such as those that the fish likely experienced in the bottom of a boat or in large earthenware crocks.

Outside China, aquatic farming for food production by the rural peasant probably developed more through primitive tribal societies that fished for subsistence and survival. In most early tribal societies, fishermen relied on a variety of fixed traps in addition to their spears and lines with hooks fashioned from bones for traditional fishing. These traps ranged from simple woven and baited baskets suspended in the water to earthen ponds constructed along the shoreline and to complex labyrinths of fences made from bamboo and covered with reeds, beguiling fish into ever-receding spaces from which there was no escape.

Much information was spread among all tribal societies through the general diffusion of mythical folklore, and the tribes that survived by fishing included fish in all their symbolic rituals. The fact that it was commonly taboo for women to work in fishponds throughout Asian and Polynesian societies as late as the seventeenth century implies that the soul of fish was well rooted in common tribal origins. Nonetheless, there is little doubt that many tribal fishermen independently discovered that most fish could be held in captivity in traps for several days to keep them fresh, and even longer if they were provided with some feed. It was to be a long time before they discovered places where seasonal resources of fry and fingerlings could be captured in large numbers and impounded in some way by damming streams and small rivers. There, the young fish were able to be grown out for future harvest.

In primitive societies, tribal hunters appear to have an affinity for their quarry. Anthropologists see this in their studies of remote communities of Indonesia and Papua, New Guinea, or Bolivia and Brazil, for example. The hunters understand the influence of the changing days and seasons on the movements of their prey and notice any effects of very local conditions. The trapping of large numbers

Figure 3.2 Bolivia, thirteenth to sixteenth century; aerial view of Middle Age fishponds and weirs.

of marine fish, such as milkfish and mullet, in coastal ponds is believed to have started in the island communities of Southeast Asia sometime between the thirteenth and fifteenth centuries. This was possible because of the ready availability of millions of young fish migrating inshore in early spring into mangrove forests rich in food. There, they were caught and transferred to shallow ponds, or *tambaks*, crudely fashioned around the estuaries and filled with sea water for the production of salt.

By the third century BC, the early history of Java was greatly influenced by both Chinese culture from the north and Indian cultures from the west, and their traditions of religion and social organization were soon well integrated into the island's many great kingdoms. There was also significant trade between the regions. The Indian epic, *Ramayana*, for example, described Java as an island rich in resources of gold, grains, spices, and particularly of salt. Consequently, it is probable that simple skills in husbanding fish in ponds were also introduced to the remote southeastern region of continental Asia at the same time as they were moving west with the Chinese through India.

The development of coastal farming in the region was probably influenced most, however, by the consolidation of Islam in India around the millennium. With the domination of India and control of trade throughout the Indian Ocean, the Moslems looked for new conquests in Southeast Asia. These expansionist policies had a significant effect on the fortitude of the different religious groups in the region, particularly the Hindus. With the increasing power of the Hindu rulers in central Java, any conquered peoples and convicts, according to the folklore of the inhabitants of eastern Java and adjoining Madura, were cast out to the distant coastal areas and neighboring islands to build salt pans and maintain fires. These outlawed unfortunates were not allowed to practice agriculture or build boats to fish and were not even allowed to wear clothes. They subsisted mostly on available fish and shellfish in the mangroves and soon found the saltwater ponds a lucrative source of available food in the monsoon seasons when the heavy rains turned them into natural ponds. They were the people who began to build the first *tambaks*, which, fortuitously, proved to be ideal fishponds for milkfish, mullet, and shrimp species that fed on the highly nutritious layers of microbenthic organisms that accumulated on the bottom. This thick, rich mat called *lab-lab* by the natives was a teeming mass of microscopic animal life fed by thick populations of green and blue-green unicellular algae, which continuously collapsed on the dead layers beneath.

Stamford Raffles, in his *History of Java* written in 1817, noted that the majority of *tambaks* in east Java were constructed in the fourteenth and fifteenth centuries under the direction of the Islamic missionaries, and later, the sultanates, but it was probably earlier than this. An ancient code of Hindu law, the *Kutara Menawa*, which originated in about 1400 and right at the end of the era, contained penalties for stealing fish from fishponds. The law also differentiated penalties between stealing from freshwater ponds (*siwakan*) and saltwater *tambaks*.

However, with the organized help of several local Islamic leaders, it was the United East India Company's desire to monopolize the trade in spices and salt in the second half of the eighteenth century that influenced the great expansion in *tambak* construction throughout all the islands. This Dutch company was formed by the *Heeren Zeventien* (Seventeen Gentlemen), who merged their independent trading enterprises into one large company in 1602. Despite a charter issued by the Netherlands government that the company could do anything in its power from waging war to concluding treaties with kings, it took another 150 years before it finally wrested control of trade from the Chinese and conquered the last Hindu kingdom. However, it could never suppress the British East India Company, which had been founded at the same time, and the two nations continued to interfere with each other's shipping all over the world for another 200 years.

By 1789, the United East India Company was bankrupt, and all Dutch overseas possessions were put under French control when the Netherlands was occupied by the French in 1795. Napoleon crowned his brother Louis the king of the Netherlands, and his appointed governor of Java made himself immediately unpopular as he tried to reform the country's mediaeval feudal system. At the same time, the British, who continued to be at war with Napoleon, tried to seize any Dutch possessions. However, with the ending of the Napoleonic Wars in Europe and the return of the Netherlands by the British to the government-in-exile, the temporary colonization of Java by the French was ended. In 1811, Stamford Raffles was appointed lieutenant governor of Java and all its dependencies by the British East India Company, and he, too, tried to change the enforced agricultural system to one of modern taxation. But before many of his comprehensive reforms were instigated, the Dutch authority over the country was reestablished. After a bitter five-year war with the remaining kingdoms in Java, the Dutch resumed total control and introduced the new Cultivation System. This act gave the Dutch complete authority over all the tropical crops grown for export and over most of the remaining agricultural lands that were then leased back to the Javanese aristocracy. Ownership of the fishponds was also transferred to the colonial government, and leased back to those who would maintain the ponds and keep them stocked with fish.

For the Javanese villagers, the Cultivation System was a disaster, but it helped to initiate Dutch expansionism throughout the entire region. A survey in 1863, one of many carried out for taxation purposes, showed that there were over thirty-two thousand hectares of *tambaks* in Java and the island of Madura for both salt and fish. New ponds were also being constructed in the western part of Java and in neighboring Sumatra through reclamation of coastal lands and estuaries. Most of the coastal ponds were leased to local Islamic leaders and subleased to farmers for a rent of money and twice-weekly delivery of fish. By the end of the century, the area was over fifty-five thousand hectares, and a survey in 1949 showed that it had expanded to some eighty-two thousand hectares, and that many of the ponds were new or rebuilt under the Japanese occupation from 1940 to 1944.

3.4   Ancient fishponds of the island societies

The spread of coastal farming did not stop in the islands of Southeast Asia. The islands of the Hawaiian chain and Tahiti, almost at the limit of Polynesian extension in the Pacific, were known to be well populated by the year 1000 from migrations of people descended from the Melanesians of the far western Pacific. Therefore, it is probable that simple skills of catching fish and keeping them in coastal *tambaks* were disseminated from Southeast Asia, through the Melanesian Islands, and out into the Pacific between the tenth and thirteenth centuries.

Records show that brackish-water ponds were an important part of Polynesian society throughout the Pacific islands at that time and that fish-holding practices became more sophisticated. The earliest recorded date for the construction of fishponds in the Hawaiian Islands is in the middle of the fifteenth century. These tidal ponds, shaped by stone and coral walls on top of reefs, were often built around or adjacent to streams so that the increased productivity of the fresh water attracted and fed young shoaling, omnivorous fish, such as mullet and milkfish.

Most of the ponds had walls with a system of one-way gates through which sea water was exchanged. These ponds were called *loko 'umeiki*. In addition to gray mullet and milkfish, many small marine fish, such as jacks, barracudas, and ten-pounders, could move in to feed and grow and were later harvested as adults as they tried to move back out to sea to spawn. This usually was coincidental with a period of full moon. Other ponds, called *loko kuapa*, had no gates at all, but instead exchanged water through the permeable coral walls.

Figure 3.3 Islands of Hawaii; ancient fish lagoon for royalty, with remaining manmade coral-rubble retaining dike in foreground (Kanoa, Molokai, Islands of Hawaii).

The ancient Polynesian walled fishponds were owned only by the chiefs, who had them built by their own village people specially drafted for the purpose. The ponds became symbols of importance to each chief. By the beginning of the nineteenth century, over two hundred ponds were recorded around the islands of Hawaii. Many were large, with walls from two thousand to five thousand feet long. Each pond had keepers who fed the fish with taro and other vegetation, and harvested fish when required by the chiefs for themselves and their immediate retinue.

The common people had no right to the fish in the ponds of the chiefs, but they were not denied their own ponds. In the sea, they made enclosures by cleverly connecting coral heads with walls, thus trapping fish as the tide receded. They were also allowed to operate fishponds inland in the wet areas, where taro was grown.

Polynesian society was marine-dependent. Many kinds of fish were abundant around the islands of the Pacific and provided the principal animal protein in the inhabitants' everyday diet. Therefore, the Polynesians were not dependent on the production of fish in the ponds, which was about two hundred kilograms per hectare, but rather, they used the ponds as a source of fish in bad weather and as a ready supply for feasts.

The last of the ponds in the Hawaiian Islands was constructed in the early part of the nineteenth century, after which their numbers steadily declined. The continuous erosion of the traditional culture and social structure in the islands removed their symbolic importance, and there was no maintenance for those destroyed by storms or volcanic eruptions. Many simply filled with silt as the lands were cleared for industrial agriculture. Several passed into private hands in the twentieth century and survived, but most of these were caught up in the subsequent demand for flat coastal land by the United States military forces and by a burgeoning tourist industry for beachfront sites and golf courses. The few that remain today are protected as cultural heritage sites.

3.5   The floodplain farms of South America

Recent archeological discoveries in the Amazon basin of Bolivia have revealed an organized infrastructure of channels and earthen ponds covering an area of more than three hundred square miles of floodplain. Radiocarbon dating of materials used in the construction of the weirs shows that the labyrinth was constructed at least in the sixteenth century and well before any possibly known influence of the Spanish conquistadors.

Prehistoric irrigation of this and other floodplains in South America with dams and an elaborate system of canals and small dikes have led anthropologists to believe that some of the early societies were not the conventional hunters and gatherers who roamed the forests and savannahs in search of birds and game. Rather, they were more complex and settled cultures, thanks to their obvious skills in environmental management. Through simple but effective hydraulic engineering, they were able to turn these tropical lowland areas, frequently noted for their poor soil as well as their heavy seasonal rains, into a combination of permanently raised fields for the production of crops, and zigzag weirs and ponds for the trapping and holding of fish. As a result of these alterations, they were creating environmental ecosystems ideal for attracting waterfowl and game that they could hunt.

Fishes characteristic of the South American floodplains, such as the colossomids, catfishes, perch, and piranhas, are well adapted to the natural cycle of wet and dry seasons that alternate every six months. The fish surviving the dry season migrate quickly and spawn profusely in response to the rising flood waters from the rains that begin around November. Most of them feed on the fruits, nuts, and seeds falling from the inundated vegetation. When the dry season comes around again in May, they are concentrated by ever-receding waters to survive on decaying vegetation, algae, and bacteria, supplemented by a diet of zooplankton, insects, and snails. Consequently, the weirs and ponds would trap and hold breeding fish, and subsequently their progeny, making this remote area of Bolivia an extraordinarily large farming complex capable of providing fish and possibly edible snails, all year round.

Some anthropologists believe these floodplain societies in South America have existed for several millennia, rather than several centuries. Such speculation is reasonable in view of the similar development of quite sophisticated irrigation systems of the Assyrians along the Tigris River and the Egyptians along the Nile before 2500 BC. Although the place of the Chinese and the other great ancient cultures of Asia in the early history of fish and shellfish husbandry is completely inviolable, it is gratifying to discover possible parallel emergence of some similar primitive practices in total isolation on the opposite side of the globe.

Bibliography

Charter of the Kladruby Monastery (1115) [Charter established the oldest documented fish pond in Bohemia, in 1115]. Cited in: Veselá, V. (2007) By bike or boat: Experience the beauty of the Czech Republic without a car. http://www.czech.cz/en/current-affairs/tourism-and-sports/by-bike-or-boat-experience-the-beauty-of-the-czech-republic-without-a-car (last accessed on July 29, 2010).

Chaucer, G. (1387 to 1392) *The Canterbury Tales* [passed down in several handwritten manuscripts], England. [Modern editions available in original Middle English with annotation, for example: Fisher, J. and Allen, M. (2005) *The Complete Canturbury Tales of Geoffrey Chaucer.* Cengage Learning, Florence, Kentucky.]

Dom Pinchon (1420) [diary]. Cited in: Fish, F.F. (1936) Founders of fish culture: European origins. *The Progressive Fish-Culturist* 3, 8–10.

King Charlemagne (812) *Capitulare de Villis.* (Die Landguterordnung [1895] *Kaiser Karls Des Grossen: Capitulare De Villis Vel Curtis Imperii,* German edn. Kessinger Publishing, Whitefish, MT.)

Kutara Menawa (1400) Javanese Law. Cited in: Janet, H. and Brown, P.B. (1987) Backyard fish farming in Java, Indonesia. *Community Development Journal* 22, 237–241.

Magna Carta Libertatum (1215) (The Great Charter of Freedoms) [in Latin]. English versions available at: http://alexpeak.com/twr/mc; http://www.statutelaw.gov.uk/content.aspx?activeTextDocId=1517519 (last accessed on August 4, 2010).

Raffles, T.S. (1817) *History of Java.* 2 vols. Black, Parbury, and Allen, London; John Murray, London.

Valmiki, a Hindu poet in India (Sixth century BC). Cited in: Keith, A.B. (1915) The date of the Ramayana. *The Journal of the Royal Asiatic Society of Great Britain and Ireland* April, 318–328. http://www.jstor.org/stable/25189319.

William the Conqueror (1086) *Domesday Book.* Gloucester, England.

Chapter 4

The Slow Dawn of Science (1450–1900)

Abstract

Construction and management of fishponds were an integral part of fifteenth century European life. Production technology progressed over the next hundred years. Fish culture techniques were published. A peak in domestication of aquatic animals occurred worldwide. Most Asian countries practiced breeding and propagation of freshwater fish; Japan improved shellfish management. The second hundred years of the Renaissance saw less interest in fish farming in Asia and Europe, and a steady decline in England after the Reformation. The Industrial Revolution in Europe and the resulting mechanization of fishing fleets in the nineteenth century revitalized and expanded marine fishing capability; modern railways carried seafood to inland markets. Industrialization resulted in uncontrolled harvesting of continental shelves, and decimated traditional estuarine and river fisheries via pollution, demand for water, and industrial use of riparian and coastal lands. Amateur and professional organizations in the new discipline of marine science began to address these problems.

4.1 Inquiring minds of the Renaissance

By the beginning of the fifteenth century, both construction and management of fishponds were becoming well-established skills and an integral part of artisan life, particularly in Central and Eastern Europe. As the word spread about the communal benefits of fish and fishing, the burghers of many towns in Eastern Europe initiated local pond construction projects. Soon, each project became a symbol of civic pride, and the towns began to compete with each other or with the neighboring noble houses and old monasteries for the best ponds.

The History of Aquaculture. By C. E. Nash. Published 2011 by Blackwell Publishing Ltd.

Throughout Bohemia and Moravia, in response to a general edict by the Emperor Charles IV, hundreds of ponds were built around the towns. Many were impressive feats of engineering and so well constructed that several have survived to this day in the guise of manmade lakes. Moreover, because of the grand scale of some of these projects, the costs of construction were very high for the times. For example, the value of a large 260-hectare pond was equal to that of two or three villages, and all their inhabitants as well as the land. In these two regions alone, the number of ponds recorded in the archives of the time covered over seventy-five thousand hectares and consequently placed a considerable demand on the natural water resources. Therefore, to conserve water, the engineers frequently designed the ponds with interconnecting canals. The most famous of these was the Golden Canal, which ran for almost fifty kilometers.

In other regions, community initiatives did not come from the burghers, but from the fishermen themselves. In neighboring Hungary, for example, the cooperation and integration of the fishing communities working the lakes and interconnecting waterways of the river plains helped to create completely new towns, such as Szeged and Komárom. Also in southern Europe, in the coastal *valli* of Italy, there were several thriving fishermen's cooperatives that became responsible for the entire economic and social welfare of their local areas.

With fish and fish culture firmly established as a respectable trade in Central and Eastern Europe, the next hundred years saw much progress in simple production technology, particularly for the carps. Most of the advances originated in Bohemia, primarily as a result of a competitive civic spirit, the great investment in large capital facilities, and a subsequent need for producing the fish to stock them. Therefore, as part of each large communal scheme for producing fish for the local marketplace, separate small ponds were built at the side of the large production ponds for holding broodstock fish, for spawning and fry rearing, and as nurseries for fingerlings. With this control over the life cycles of the fish and propagation of the seed, however tenuous, management of the production ponds became more and more intensive. Yields increased steadily, and records show that seventy-five to one hundred kilograms per hectare was the norm.

Bohemian engineers became famous and were much in demand throughout Europe for advice on the construction of canals and ponds, as well as for information on fish culture. One such man was Stepanek Netolicky, who traveled widely designing and managing the construction of pond systems and giving advice on culture principles. Netolicky was probably the first fish farming consultant, over four hundred years ahead of his time. His successor was Jakub Krcin, another hydraulic engineer famous for his ambitious schemes of creating new waterways and constructing large ponds, many of which were over three hundred hectares in area and more than ten meters deep.

It was during this period that the early techniques for intensified fish culture began to appear in published books. In 1547, Ioannes (or Janus) Dubravius, the bishop of Olmutz and a contemporary of Netolicky, produced his Latin treatise, *Jani Dubravil de piscinis et piscium qui in illis aluntur libri quinquae*. The book,

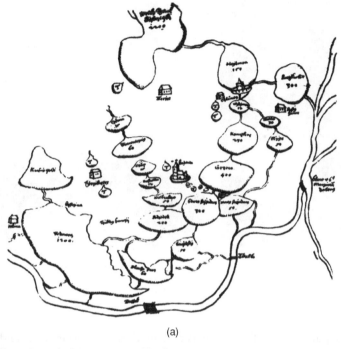

(a)

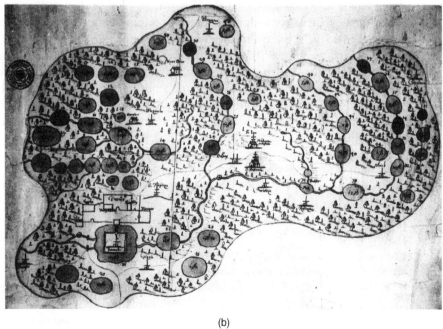

(b)

Figure 4.1 Southern Bohemia; sixteenth to seventeenth century fishpond maps: (a) 1520 fishpond system of Trebon; (b) 1657 fishponds belonging to Cervena Lhotka Castle. (Courtesy of Rudolf Berka.)

(a)   (b)

Figure 4.2 Southern Bohemia; modern fishmaster and pond fishing: (a) fishmaster of Pond Rozmberk in 1929 wearing the traditional dress of the guild; (b) pond fishing in the Trebon district. (Courtesy of Rudolf Berka.)

first published in Breslau, was reprinted in Latin four times and translated into many languages, including English, German, and Polish. It was very advanced for its time, particularly with its chapters on economics and diseases. Dubravius also described in great detail a new technique for carp breeding: that is, he advised preparation of small ponds that were overgrown before they were filled with water. A pair of carp broodstock was to be released into each prepared pond and removed after spawning, leaving a productive and manageable habitat for the progeny until they were old enough to be transferred to a nursery pond. Finally, he faithfully described the work of the fishpond wardens and their organizational hierarchy.

Other books on fish culture slowly began to follow. In England, John Taverner published his treatise on *Certain Experiments Concerning Fish and Fruites* in 1600, followed by Gervais Markham with *Cheape and Good Husbandry* in 1623. Both described the practices of raising fish, and particularly carp, in considerable detail. These two works were so similar in content that one might suspect some plagiarism. Each not only described a suitable layout for a small production farm of about five hectares, but also gave advice on individual pond construction, pond fertility, fish management, feeding, and the best fish to raise. They also made acute observations on the breeding behavior of the fish and on the care of fry and fingerlings. Finally, both explained the benefits of allowing terrestrial animals to graze on ponds dried in rotation, leaving rich manure to fertilize the soil for the aquatic vegetation, which would grow back quickly once the pond was refilled. They also noted that the manure could be a direct source of feed for the fish.

The beginning of the Renaissance period coincided with the first peak in the early domestication of aquatic animals worldwide. By this time, most countries

(a)

(b)

(c)

Figure 4.3 Bohemia, sixteenth century; Ioannes Dubravius: (a) etching of Ioannes Dubravius, Bishop of Olmutz; (b) cover of his book on fish culture written in 1547; (c) Dubravius' headstone, on which he was remembered through the image of a fish for his contribution to aquaculture. (Courtesy of Rudolf Berka.)

of Asia practiced breeding and propagation of several freshwater fish in captivity and had developed relatively sophisticated production practices, based on the original skills of the Chinese that were spread throughout the region.

The Japanese, too, were becoming more adept in the management of their sessile shellfish, such as clams and oysters. Advances began quite indirectly through the help of the great clan lords, the *Shogun*, who were wont to move around the countryside with their armies, complete with large entourages carrying all their necessary equipment and food supplies. Their preferred menus included shellfish, especially clams and oysters, which the bearers transferred from place to place and then stored back in the sea. The coastal fishermen of the Inland Sea observed how readily the transplanted oysters spawned, no doubt because of the temperature shock. Soon, the bamboo fencing surrounding their clam beds and fish traps were covered in young spat. From there, the fishermen began to develop the culture of oysters on bamboo poles deliberately stuck upright in the sand and mud. For the most part, this ancient system is still evident today.

At the same time, in the nearby island communities of the South China Sea and far out the Pacific Ocean, the village menfolk were actively raising marine fish. The fish could not breed in the ponds, but the fishermen gathered them in large numbers as they came into the coastal mangrove forests to feed, and they stocked them in the ponds they constructed on the shoreline from coral and rock.

Similarly, in countries of the Old World, methods for husbanding freshwater fish in captivity were steadily advanced as the practice was being transferred from the hands of the monks and nobles to the village commoners. However, they too had to rely on natural breeding to supplement their stocks.

The second hundred years of the Renaissance saw little further gain of any substance. With the Reformation in England, there was evidence of a steady decline in subsistence fisheries and fish farming. First, the dissolution of the monasteries in Britain by Henry VIII in 1541, and then the Protestant revolution manipulated by the protectorship of the young Edward VI in 1547 saw the closure and physical destruction of many monastic lands and confiscation of church properties, together with their well-engineered water channels and stew ponds. Henry even imposed fines and penalties on anyone caught buying fish from foreigners so that he could keep up the pressure on the Catholics and their plethora of religious feasts. With the sudden economic hardship in the country, and without the leadership of the clergy, many people lost their means of subsistence. They became less and less observant about fast days, even during Lent. However, there was an unfortunate twist for the ambitious factions within the protectorship.

Without the regular demand for fish by a large part of the population, the bountiful offshore fisheries between England and Europe suddenly became dominated by the Dutch. Furthermore, there was a sudden loss of interest in building any more seaworthy vessels. It took the foreign policies of the newly crowned Queen Elizabeth I to stop the decline. In 1562, she declared that every Wednesday was "to be used and observed as a fish-day," and disobedience was subject to fines and other penalties. The statute stated her clear intent, that it was "meant politically for the increase of fishermen and mariners, and repairing of port towns and navigation, and not for any superstition to be maintained in the choice of meats." Getting the Dutch out of the fisheries subsequently contributed to the protracted naval scuffles between the two countries, which would continue off and on for some two hundred years.

With the growth of seaborne trade of most countries in the Old World through the years of the Renaissance, there was increasing competition throughout Europe for water and land. The competition was not necessarily confined to areas around coastal ports and trading centers; it also existed far inland. It began slowly, with simple competing demands for well-drained, fertile pastures to meet the needs of an improving agriculture industry, particularly for the emerging systems of intensive crop and animal husbandry. A century later, the competition had become a rout, as the Industrial Revolution made insatiable demands for all the different uses of these two resources.

Throughout the Renaissance period, a similar sudden lack of interest in fish farming was evident in Asia. It started right at the outset, in 1433, with the Great Withdrawal of China. As abruptly as she had started, the influential giant suddenly stopped showing her hands filled with achievements new and wonderful to the rest of the world. She slammed her doors as she tried to deal with the invading Mongols from the north, and nothing more would be seen of Chinese creativity until the arrival of the missionaries and soldiers of fortune from the Western world.

4.2   The Industrial Revolution in Europe and the mechanization of the fishing fleets

The nineteenth century is looked upon in history as golden years for science and technology. It was a time of individual discovery and enterprise, which created an excitement in the Old World that will never be repeated. Many industries made great advances during that time, and the useful but relatively unimportant industry of fishing was no exception, particularly in Europe, the Scandinavian countries, and across the Atlantic Ocean in North America.

At the dawn of the century, the demand for fish and shellfish was met adequately by the traditional coastal and estuarine fisheries, and some dominant inland fisheries. Although the Napoleonic Wars in Europe would leave the fisheries in a woeful state, with fleets once again run down and fishermen sufficiently impoverished to leave for work in the mines and factories, the general distribution of resources fitted admirably the distribution of the relatively sparse population at that time. People were widely dispersed among rural villages or grouped in small agricultural towns and a few principal cities that were trading centers. By the end of the nineteenth century, the demand was met by highly productive, expanding high-seas fisheries. Fishing was almost fully mechanized, and modern railways carried the fish to inland markets in industrial towns and new cities. The inland fisheries had all but disappeared.

A number of singular events collectively had a dynamic impact on the growth of marine fishing. These would transform it from a meager way of life for those who lived by the sea to a major industrial economy. New fishing grounds were being discovered by groups of fishermen who were willing to move and to follow the shoals of fish on their seasonal migrations. The more distant offshore sites were worked more efficiently by use of a combination of vessels: a trawler remained on the fishing grounds for a week or more, and a fast cutter ferried the catch back to one of a number of ports, depending on the prevailing winds. In this way, the most enterprising fishermen exploited the rich resources of fishing banks further and further afield.

Middle-distance fishing, as it was called, was also made more practical by the availability of larger quantities of ice to preserve the catch during the run for port. Machine-made ice was already replacing the use of natural ice. The Scandinavian idea of using ice for shipping and preserving fresh produce had been adopted for preserving fish at sea. Fishing was suddenly made very profitable as the increased catches of high quality fish found large and ready markets to absorb them. The great labyrinth of railway networks for the new high-pressure steam engines provided fast links from the fishing ports to the large industrial inland towns. Cheap, fresh fish was therefore no longer the prerogative of only the fishermen and the inhabitants of small coastal towns, but rather, it became available to everyone within reasonable reach of the sea.

The advantages of the new steam-driven engine for fishing vessels were not lost on the marine engineers of the day. Steam-driven vessels were first used to take the catch from the sailing trawlers back to port and to tow becalmed vessels. Next, they were used experimentally to pull sailing trawlers as they fished with the heavy beam trawl. Then, in 1854, the first fishing boats were designed for both sail and steam, and they were designed jointly by three railway companies that had eyed the potential for their new investments in steam transportation. However, it was not until 1880 that the first effective steam trawler was built and tested. It was an instant success. Sailing fleets were rapidly replaced by the new steam-propelled vessels. As a result, trawlers in ever-increasing numbers ventured beyond the middle-distance grounds to fish at sites even further afield.

If all these industrial advances were not enough, the fishing equipment was also mechanized, greatly improving the operational harvests at sea. Beam trawling for fish from boats under sail already had been practiced for two centuries or more when the otter trawl was introduced in 1894. This innovation was quickly adopted by the fishermen, because with its two outriding boards or doors, the otter trawl was easier to handle than the heavy and cumbersome beam trawl. It also used less deck space for manipulation and storage. Moreover, the double-barrel winch was also driven by steam, rapidly speeding up each haul and minimizing the time it took to fill the boat with fish. All these advantages were accepted gladly by the design-conscious marine engineers, who were looking for even greater range and efficiency in their new vessels.

The design of the otter trawl modernized fishing methods and completed the mechanization of the fishing industry in a brief period of fifty years. Industrialization of fishing brought about an explosion in uncontrolled harvesting of all continental shelves within two to three hundred miles, and all the fertile banks of the North American coastline and its neighboring seas were easy targets of overexploitation.

With all this exciting technical development, most countries with a maritime coast had an interest in taking economic advantage of the available resources. Many immediately conducted public inquiries on the state of their fishing industries. In Great Britain, for example, the government appointed a royal commission in 1863 to examine the state of British fisheries, and yet another commission fifteen years later to do it again. Marine fisheries and fishing became so interesting that two international exhibitions were organized: one in Edinburgh in 1882, and the next, even bigger, in London in 1883.

The sudden expansion in fishing capabilities of coastal nations with large mechanized fleets and modern rapacious gear was potentially a serious problem. Because most of the rich and newly accessible fishing banks were also the breeding grounds and nurseries of marine fish, there was a risk that overfishing and the lack of regulation of the mesh size would wipe out the fry and juveniles recruiting into the fisheries. Moreover, the effects of pollution from oil-driven vessels were not going unobserved by the ornithologists and naturalists who frequented the beaches and estuaries, or by the coastal oystermen.

Fortunately, these growing problems in European waters and their possible consequences were recognized by some of the leading public figures of the time. Through their awareness, they formed organizations to press for regulation and management of the new fisheries. Their concerns were taken seriously, because among their number were distinguished scientists of the Royal Society who pushed for the collection of factual information and research in this new field. They were also backed by many brilliant amateur naturalists who were willing to provide their own money and effort, as necessary, together with any political influence they might carry.

Thus, fortuitously, in parallel with the birth and development of the modern fishing industry in the nineteenth century, the new discipline of marine science emerged. Among its many fields, it included fisheries biology. The professional recognition of marine science was helped by the flourishing Linnaean Society, which was by then almost one hundred years old. Since its inception, the prestigious Linnaean Society had always encouraged each new group of scientists and naturalists to form professional groups of their own, under the banner of their own discipline. This profound advice helped give birth to the Challenger Society, named after the famous world voyage of HMS Challenger from 1872 to 1876 to study the science of the seas, and the International Council for the Study of the Sea in 1902. Further, the Europeans were not alone. The same awareness and recognition of the problems were also apparent in the New World, and in 1872, there was the first meeting of the American Fish Culturists Association. This small association in later years would change its name to the American Fisheries Society.

4.3   The demise of the inland fisheries

At the same time as the first wave of marine and fisheries scientists and amateur naturalists were preparing and organizing themselves to meet the challenge of uncontrolled developments in the sea fisheries, they also became unwilling witnesses to the changes and uncontrolled developments in the inland and estuarine fisheries. The golden years of marine fishing were about to become the dark ages of the freshwater fisheries. During the exact period when the coastal fisheries were taking advantage of the Industrial Revolution to expand at unprecedented speed, providing people with fresh nutritious seafood, the same revolution was rapidly decimating the traditional estuarine and river fisheries that had served the Old World dependably for so long.

The traditional inland fisheries of the countries of Europe and the North Atlantic rim were based on migratory species, predominantly the Atlantic salmon and to a lesser extent, the eel. However, in the hundred years between 1750 and 1850, the demand for water by innumerable new industries and towns that grew on the banks of every river and fast-flowing stream drained and polluted the once pristine breeding grounds and runs of the salmon beyond recall. The effluent of untreated sewage from towns and the tail-washes from chemical and industrial factories were the main causes of pollution and subsequent decimation of the inland fisheries and damage to the shellfish fisheries in the nutrient-rich estuaries along the coast. Large volumes of fresh water were suddenly removed for energy to drive the factory machinery and to fill the vast networks of canals that were being built to transport the raw materials and finished products between the factories and the coastal ports. Consequently, rivers were dammed, diverted, or totally rechanneled, preventing passage of salmon on their annual migrations between breeding grounds and feeding grounds.

In addition to these environmental pressures on the fish, rivers and streams were by that time heavily fished by all within their reach. Catching a live fish could have been a bright moment in the monotonous lives of lowly paid workmen in the oppressive factories of the new industrial towns. They took advantage of fishing in local waters as they struggled for survival. It was not always a sporting match, because poachers used every conceivable device, including fixed and floating traps, illegal nets, explosives, and poisons in their unstinting efforts to get a free meal for the family. Although laws had been on European statutes since the fourteenth century that recognized ownership of salmon and its waterways, prohibited fixed engines and barriers, and ordered closed and open seasons for harvest, the existing regulations were not enforced by the authorities and were consequently ignored by industrialists and their workers alike.

Internal industrial growth was almost entirely responsible for the changes that brought about the demise of the traditional river and estuarine fisheries of Western Europe, but it was not the only cause thereof in other locations. Deleterious changes were also taking place among the continental fisheries, particularly in the extensive inland lakes and watershed ponds in countries of Central and Eastern Europe. Since the Middle Ages, there had been a considerable development of the fisheries of the common carp. Although fish culture still resembled an artisan craft, the business of managing ponds and producing fish had been organized and structured like that of many other more important tradesmen, such as carpenters, coopers, and silversmiths. Just like the other guilds of merchants that had their roots in the Middle Ages, the national Fishermen's Guilds were formed to keep tight control on the way things were done, and to set high standards of training and qualification. The members even wore formal dress at work. However, with new developments that were occurring in agriculture, and the interest in more land for animal husbandry and crops, many of the low-lying lands that had been ideal for fishponds were drained and lost. In Bohemia, for example, the area under fish cultivation for carp and other freshwater fish, such as tench, pike, pike-perch, and sheat-fish, had been more than halved by the end of the century to create new and fertile pastures for raising cattle and sheep.

By the time the European parliamentarians finally recognized the many problems facing the once-productive inland fisheries of their continent and took action with a number of national laws and regulations to protect natural waterways and migratory fish, the damage had already been done. All the large and most productive rivers in Europe, which had supplied fish to eat and a living for fishermen for centuries, were polluted and impassable to salmon, and to this day salmon have yet to return. It was not until the end of the nineteenth century that the need for scientific information and biological research was fully recognized, if the damage to the inland fisheries was to be stopped and rectified. Once again it would be due to the awareness of the small group of influential scientists and amateurs of the time, and the formation of both public and private organizations to safeguard and monitor the inland as well as the marine fisheries and to undertake scientific research to bring about their rescue and recovery.

Bibliography

Dubravius, I. (or J.) (1547) *Jani Dubravil de piscinis et piscium qui in illis aluntur libri quinquae*. Breslau, Poland.

Markham, G. (1623) *Cheape and Good Husbandry for the Well-Ordering of all Beasts and Fowles, and for the Generall Cure of their Diseases*, 3rd edn. [plan of fish ponds], p. 174. Thomas Harper for John Harrison, London.

Taverner, J. (1600) *Certain Experiments Concerning Fish and Fruites*. W. Ponsonby, London.

Chapter 5

The Roots of Modern Aquaculture (1750–1880)

Abstract

Seaweed farming originated four hundred years ago in Asia, after millennia of wild harvest. Natural history study blossomed in the nineteenth century, with interest in classification, selective breeding, and evolution. After a French team's artificial fertilization of trout eggs, the first fish hatchery at Huningue (1852) distributed eggs to European rivers. Freshwater hatcheries proliferated. Buckland created wooden-box incubators, publishing methods in Britain. In Bohemia, Horák produced a fish farm management and fish culture handbook. Danes were first to feed cultured fish in captivity instead of releasing to streams—the first European land-based fish farming. Similar experiments succeeded in North America. Hatcheries were built, and fry transplanted from coast to coast. Common carp introduced from Germany to California in 1872 flourished. From America, fertilized fish eggs were shipped worldwide. Introduction of cultured fish benefited sportsmen and fish managers, and helped compensate for population loss due to damaged spawning areas and overfishing.

5.1 The farming of seaweeds in Asia

The mantle for developing the crude technologies for culturing aquatic animals and plants in Asia became available when China renewed her years of silence in the middle of the second millennium. Slowly but surely it was taken up by the Japanese, who would in time find it to be a perfect fit. It began with the farming of seaweeds.

Seaweed farming has no seeds in antiquity. Although Asia is once again recognized as its birthplace, by comparison with the ancient beginnings of fish

---

The History of Aquaculture. By C. E. Nash. Published 2011 by Blackwell Publishing Ltd.

farming, the event occurred just yesterday. Seaweed farming only originated within the last four hundred years. This implies that for over four thousand years, the economic uses of seaweeds, of which there were many, relied on hunting and harvesting practices that managed and sustained the abundant natural beds.

Unquestionably, aquatic flora had been used by primitive societies for a variety of purposes, and especially the green tender leaves of freshwater plants and algae. The plants of the coastal shorelines, in contrast, must have presented a formidable challenge at first, with their tough leathery thalli and thick rubbery stalks. However, for the persistent foragers and collectors, the bright green or red leaves and filaments of the more tender varieties found in the intertidal zones would have been discovered before too long.

The first recorded use of seaweeds was described in Chinese literature. As long ago as 3000 BC, the Chinese were harvesting seaweed for medicinal use, and Shen Nung, the "father of medicine," prescribed seaweed for a number of ailments. Confucius in about 500 BC referred in his poetry to both the medical and nutritive virtues of seaweed. But the most complete description of the varied uses of agar extracted from the fronds of seaweeds was written by Chi Han in AD 300. Seaweeds were also the basis of several popular drinks. Indeed, one such commodity was so highly valued by the Chinese that it was used as a ceremonial offering.

Despite the great esteem of seaweeds in China, the difficult task of harvesting seems to have been left for children, who were rewarded for their pains with pieces of seaweed jelly. The work would not have been easy or productive, because the long, exposed shorelines of China, compared with the sheltered rocky bays of Japan, were less suitable for large beds of algae, and the intertidal flora was less diverse. Hence, historic credit for exploiting the natural resources of seaweeds for human food and managing the beds as they did land crops goes to the Japanese. They were responsible for discovering many uses for the different gels in foods, such as noodles and soups, and for identifying the many species that could be eaten raw as a salad or preserved in brines.

In the Western world, it was left to the far-flung Celtic tribes to leave some evidence of the usefulness of seaweed. Pinned along the western-most coasts from northern Spain to outer islands of northern Scotland by the warring hordes on continental Europe, the Celts became very familiar with seafood as yet another course in their prolonged and frequent feasts. The Celts living in Wales, close to the southern shores where red seaweed called laver grew in abundance, prized a type of black bread they made from the boiled seaweed fronds.

Despite the relative novelty of farming seaweed, the date of its origin is not clear. Jesuit missionaries, who first came to Japan in the sixteenth century, described crude collection practices of seaweed. Some Japanese historians say that the farming of seaweed started in Hiroshima Bay, in the Inland Sea of Japan, toward the end of the seventeenth century. Others say quite definitely that it began in Tokyo Bay in 1736, and that seaweed was cultivated only in that one part of the country until 1818. Then it began to spread up the coast to the sheltered bays in the province of Chiba and down the coast to Shijuoka and Aichi. During the Meiji years, the technology for culturing seaweed was heavily promoted by the enlightened government in all the provinces that had a suitable coastline.

The technology was relatively primitive, but it worked. Capitalizing on the observed seasonal life-cycle of the plant, and the colonization of laver (known as nori in Japanese) on the wooden traps and ropes set out in the bay to catch fish, the fishermen dug in well-branched trees or long bamboo poles to which they attached bundles of brushwood. These devices were called *hibi*. When planted in the intertidal zone each autumn, the *hibi* proved to be efficient collectors, and soon the twigs and brushwood were filled with the spores that would grow into tender plants. By the end of the year, harvesting could begin and would continue until the spring.

Through the encouragement of the Meiji government, the early researchers began to replace the traditional *hibi* with coarse hemp ropes and then nets. In addition to being roughly textured for easy attachment of the spores, the hemp ropes also retained moisture that protected the young plants during low tides. As a result, the new surfaces increased survival, which in turn increased overall seaweed production in Japan. However, it would be many more decades before further advances in science and technology would develop the industry again.

5.2   New developments in the Old World

The artificial culture of fish to replenish falling stocks devastated by the Industrial Revolution in Europe was one small part of the regulatory response for the management of both marine and inland fisheries. But it was not the only reason for its development in the nineteenth century. The eighteenth century Age of Reason had been a time of great challenge to traditional thought in Europe, stimulating intellectual and scientific curiosity in anything and everything.

Although discoveries in the new fields of physics and chemistry evoked the greatest challenges through experimental research in university laboratories, natural history was not overlooked—principally because it could be practiced by almost anyone with an inquiring mind, and at very little cost. It just required time. Careful observation, meticulously annotated in diaries and frequently accompanied by wonderfully accurate yet artistic drawings, was the principal tool that satisfied the curiosity of the early naturalists who were eager to classify all plants and animals in the systems developed by the Swedish naturalist Linnaeus (Carl Von Linné) and the Frenchman, George Louis Leclerc, Comte de Buffon. Moreover, these were basic skills taught at most of the great gymnasiums and academies of Europe, which quickly embraced the natural sciences with the classics.

In the middle of the nineteenth century, budding biologists and naturalists of the day were given further encouragement by two important and complementary events. The first was the publication in 1859 of Charles Darwin's *The Origin of Species,* which pronounced his conclusion—after twenty years of research that took him all over the world—that all animals and plants evolved through adaptation to their particular environments. The second, in 1866, was the publication of the lifetime work of Gregor Mendel, the Moravian Abbot who, while managing the vegetable garden of his monastery at Brünn, discovered the patterns of simple heredity from the modest green pea plant.

Both of these events caused a substantial stir among scientists and opened new doors in the field of biology, evoking great enthusiasm. They also caused a substantial stir among the general public, and the fame of Darwin and his voyages of discovery on HMS Beagle whetted their curiosity about nature. At first, their curiosity was satisfied by visits to municipal museums, which displayed tropical settings filled with stuffed and preserved specimens brought back by Darwin, Alfred Wallace, and other collectors on their voyages of discovery. But then the capital cities of Europe became more adventurous, and the city fathers commissioned zoological and botanical gardens for keeping living plants and animals to provide a national attraction for visitors.

Not to be outdone, many small towns on the coast quickly joined in. With the help of the steam railway networks spreading across every country, seaside towns were becoming popular resorts for the annual holidays of the factory workers. To provide an interesting setting for their offer of bracing sea air and holiday entertainment, the town councilors ordered the building of promenades along the seafront interspersed with piers to carry the public out over the sea. Then they added aquariums that kept fish and shellfish alive in glass-fronted tanks filled with sea water. It was from such early attempts to keep marine creatures alive in tanks that city engineers would experience firsthand all the difficulties of handling salt water, not only the corrosive effects on the cast-iron pumps and pipes, but also the effects of heavy-metal poisoning on the animals and plants. These were two problems that would beset future fish culturists for decades.

The lethargic behavior of captive creatures in simulated manmade environments was of little consequence to the masses of holiday-makers, who flocked to the aquariums and zoological gardens. However, it had never been good enough for the early naturalists for whom observation of the behavior of wild animals and birds in their natural surroundings had always been the great challenge. Especially for those naturalists eager to study the behavior of fish and other aquatic creatures, water presented an added dimension of difficulty.

One of the earliest known pioneers who managed to circumvent the problem of underwater observation was Ludwig Jacobi, the owner of a large estate at Hauhenhausen in Hanover. Jacobi was a lieutenant officer in the army, but he must have had time to spare, because he applied his observations on fish behavior and life history in nature for over thirty years of his life into an accurately simulated technique for artificial fertilization and rearing of fish. Since 1740, he had patiently studied the trout and salmon in the waters on his estate, and he described his work in an article that was published in the *Hanover Magazine* in 1763. Jacobi constructed a simple incubator box fitted with grills, which he partially filled with gravel. In it he placed ripe eggs of the trout that he had carefully fertilized with the milt of running males, and put the box in a gently moving stream of water. He accurately observed the time of incubation and noted the influence of temperature on development. After incubating the eggs, he reared the young fish in small reservoirs of pure water. He continued his pioneering work for many years with the help of his sons, and in 1771, he was rewarded by another Hanoverian, the young King George III of England, with a life pension in recognition of his lifetime's work.

Jacobi's 1763 publication was written in German, and although he sent it to a number of French workers, it was seemingly ignored. However, once he began to receive public attention, particularly from royalty, his paper was subsequently translated into French in 1773 and included in a general treatise on fishes by Duhamel du Monceau. His work was also the basis of information on the successful breeding of the common trout described at some length by William Yarrell in his *History of the British Fishes*, published in 1836.

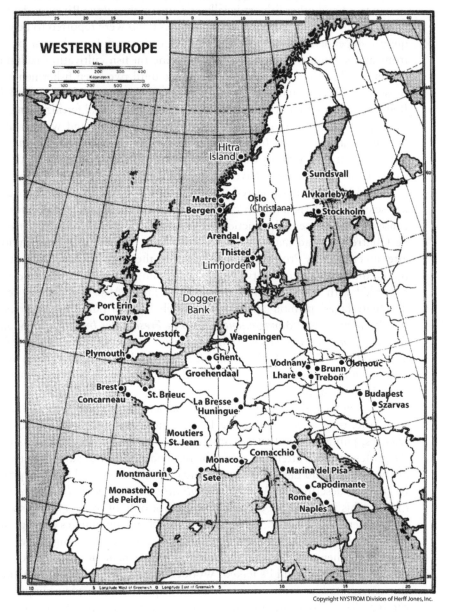

Copyright NYSTROM Division of Herff Jones, Inc.

Figure 5.1 Western Europe. (Adapted from basic map, copyright NYSTROM Division of Herff Jones, Inc.)

Jacobi was not alone. Lazare Spallanzi, a monk and professor of physiology at the University of Pavie in Italy, described the artificial fertilization of fish eggs, but the origin of his work printed in 1768 is uncertain. A few years later, in 1773, Johann Gottlieb Von Schönfeld published a book on the foddering of fishponds, following his own experiments and acute observations on the biological productivity of carp ponds.

By 1823, it was known that the fish farmers of Bohemia were attempting to spawn trout and other fish artificially at a site on the Otava River, and a number of the fishing gillies on the large estates in Scotland were reported to be working quite independently on the fertilization and successful incubation of eggs of trout and salmon. However, nothing specific on fish cultivation techniques is documented in the early scientific literature until two keen fishermen from La Bresse, near Remiremont in France—Anton Géhin, an innkeeper, and Joseph Remy, a farmer—first described their work on the artificial fertilization of trout. Their paper, called *Fécondation artificielle des poissons* and written in 1844, was not to appear in the *Journal des Traveaux de l'Académie Nationale à Paris* until 1851. It told not only of the successful fertilization, but also of attempts to stock the young fish in the River Moselotte and to raise them in a pond.

In 1848, a memoir appeared from Armand de Quatrefages to the Académie des sciences in Paris that fish culture was capable of compensating for all the causes of the present destruction of fish eggs. It was believed by many that de Quatrefages had made the initial discovery, but the Académie announced after an enquiry by Professor Milne-Edwards, the famous British naturalist, that the credit was due to the two fishermen from La Bresse, who were suitably rewarded with honors, state pensions, and fame. Also in the same year that de Quatrefages' memoir appeared, Gottlieb Boccius published *A Treatise on the Production and Management of Fish in Freshwater by Artificial Spawning, Breeding, and Rearing.* His work had a sinister message in the subtitle—*Showing also the Cause of the Depletion of all Rivers and Streams.* Boccius worked on salmon and trout fertilization in England, and indicated in his book that he had been doing so since 1815.

Other publications rapidly followed, particularly in France. In 1853, Professor Coste, then at the Collège de France and a member of the Académie des sciences, published his *Instructions Pratiques sur la Pisciculture.* This brought the professor much attention, which he relished, and he was subsequently invited all over France and Europe to lecture and give advice on local fisheries problems. In 1861, he described his *Voyage d'Exploration sur le Littoral de la France et de l'Italie.* This included the artificial culture not only of oysters, but also of eels. After observing the methods of collecting oyster spat on bundles of twigs in Italy, the enthusiastic professor set about revitalizing the diminishing oyster beds of France. He then made several innovative improvements for artificial collectors, off which the spat could be flaked. Wooden planks covered with pitch and resin that constituted his first attempts were subsequently replaced by simple ceramic roofing tiles covered with lime and sand. His successful recommendations for the reorganization of the oyster industry on the west coast of France attracted the attention of Emperor Napoleon III, who decreed the creation of two Imperial

Oyster Parks in the Bay of Arcachon. Arcachon soon became the principal center of oyster production in France, and Coste was generally accredited with saving the oyster fisheries of the country.

The ubiquitous Coste was never recognized in the literature with a forename. He was mostly referred to as M. Coste, probably the French abbreviation for *Monsieur*. In papers and letters recorded in the *Annual Report* and the *Bulletin* of the U.S. Commission of Fish and Fisheries (commonly known as U.S. Fish Commission), he was very frequently given the initial *P*, but sometimes *B* and *C*. He himself never gave the secret away, and in his reports to his emperor in 1851 on the situation of oyster production in France and Italy, he signed himself, *Your obedient servant, COSTE*. Nonetheless, it was the same M. Coste who was responsible for founding the first fish hatchery, which was built by the French government at Huningue, near Basle, in 1852. From there, eggs were distributed to many rivers across sixty-three French departments, and it was from the Huningue hatchery that eggs were first sent to intended fish culturists in England and to ten other European and foreign countries before they had their own hatcheries built.

The hatchery at Huningue was an impressive structure designed by Coste and engineered by Monsieur Coumes, the chief engineer of the Rhine and director of pisciculture. The main buildings had all the attributes of a modern hatchery, with small incubating units and long hatching troughs, and large tanks for the fingerlings. Eggs from a number of species were collected from Switzerland, Germany, and other European countries and brought to Huningue for incubation. The hatchery operated well for species that spawned in winter, but not for those that spawned in spring and summer months. After many futile years of effort, the *Messieurs* Coste and Coumes had to be satisfied with the culture of less important species, such as the lake herring. In many ways, this was a great disappointment during a time when the hatchery tried to establish itself on a commercial footing and sell eggs to prospective clients from other countries. Nonetheless, the list of species produced at Huningue was impressive. Rhine salmon (Atlantic salmon), trout, *ombre chevalier* (char), fera (lake herring), heuch (Danube salmon), alose, sturgeon, sterlet, and silure (wels) were all raised from eggs and transferred out to rivers with seemingly high survival (over 80%) in transit.

The Huningue hatchery obviously set a precedent for delays and cost overruns, two common characteristic of all fish and shellfish hatcheries that would be built in the late twentieth century. The original estimate for the Huningue facility was 154,000 francs, but when it was finally finished in 1862, it had cost the French government exactly 261,186 francs. But the hatchery was far from utilitarian. The main building was constructed with observation galleries to accommodate the many distinguished visitors who flocked to see the phenomenon of fish culture. The facility was so impressive that many visitors to Huningue immediately returned home and replicated its basic concepts, and soon freshwater fish hatcheries appeared on rivers from Western to Eastern Europe.

No country appeared to be left out. Hatcheries even began to appear in Finland and Russia, despite their natural wealth of fish. In Finland, the way was led by a man called Holmberg, who had been specifically appointed the

View of the Huningue Establishment

(a)

Basins for the young fish

(b)

Figure 5.2 France, 1862; Huningue Hatchery Station: etchings of (a) the Huningue Hatchery Station; (b) tank room for young fish in the interior; (*Continued*)

The Huningue Fish Culture Establishment

(c)

Figure 5.2 (c) egg-hatching room, showing egg trays and water system.

inspector of pisciculture by the Bureau of Agriculture of the Imperial Russian Senate for Finland. He apparently was directly responsible for hatcheries at Stokfors, Souttè, Tammerfors, and Aborfors, and had his hand in about five more. Holmberg and his colleagues were mainly interested in salmon, trout, and whitefish. In Imperial Russia, the Department of Agriculture and Rural Industry had built the Nikolsk hatchery, about three hundred miles by railroad and stage-coach from St. Petersburg, also to produce whitefish, perch, trout, and salmon. The hatchery was built on the estate of a wealthy landowner called Vrassky. Rich dilettantes were an important part of scientific enlightenment throughout Europe in those days and were welcome in most fields of study. Vrassky had begun to experiment on his properties in the 1850s with the breeding of eel-pout, trout, and perch. Later, by careful observations and the help of an embryologist called Knoch, the two discovered a "dry method" for fertilizing eggs of river-perch with the sperm of the species. Finding the method similarly successful for all of the species, he obtained the support of the Imperial Government to build a very large fish culture station on his estate, with water fed from the Pestooka River. The station was modeled after Huningue, but was extremely large, with a hatchery for over two million eggs and a variety of ponds of all sizes to produce fod-der fish, such as young bream, carp, and other whitefish. Later the government built another station at Suwalki, also with the motive of raising revenues from the increased fisheries.

Another visitor greatly impressed by the Huningue hatchery was Lord Gray, a wealthy landowner in Scotland and a proprietor of the salmon fishery on the famous River Tay. On his return home, he persuaded the rest of the partners in the private fishery to build a similar hatchery on the Tay. Faced with the evidence of declining fish runs, they readily agreed. A small facility was built at Stormontfield, near Perth, and top fish culturists of the day were hired to carry out the program. The hatchery proved to be a success, judged by the marked returns of salmon to the river, fully satisfying the proprietors of the river, the holders of the fishing leases, and presumably, the anglers.

After Stormontfield, two more freshwater hatcheries were built in Scotland in about 1873. One was constructed at New Abbey, near Dumfries, to restock the fishery of the River Solway, the major salmon river forming the border with England. The other was inland at Howietown, near Stirling.

In 1861, Frank Buckland began to work on the culture of a number of freshwater fish species using simple wooden boxes for incubators, and he successfully reared perch. His purpose, always, was to use fish culture to enhance the natural fisheries, rather than to produce fish independently for food. Because of his ability to write persuasively, Buckland soon became the leading authority on the new technology of pisciculture. He gave his first lecture about his experiments on fish culture to the Royal Institution in April 1863, and subsequently used the material for the basis of his book called *Fish Hatching*, which became greatly in demand.

Buckland also broadened his interest into oyster culture, following his extensive survey of the oyster fisheries in Europe in 1867. It was not surprising that in 1867, following the recommendations of the Royal Sea Fisheries Commission in 1866, he was asked to accept the position and title of National Inspector of Salmon Fisheries, one of many positions that confirmed him as the leading authority on fisheries management until his death in 1880. Thus, he set about trying to repair the damage that was being done to the rivers and estuaries by explaining the problems and working with the industrialists and fishermen to correct them, rather than by directly opposing them. It was his knowledge and concern for the development of the young fish that enabled him to obtain their empathy, although in addition, several important laws for the protection of salmon were enacted during his tenure.

Interest in fish and shellfish culture was not only centered around Coste and Buckland. There were many others who had been making the same practical experiments in some quiet corner of their homeland, because more memoirs and books appeared throughout the 1860s. Although the industry was well into a decline, the real strength of knowledge still rested among the traditional fishermen's guilds of Bohemia, and leadership still rested with the Horák family, which had maintained the Schwarzenberg Fish Farms for over three hundred years. Václav Horák produced a simple handbook on fish and fish farm management for use on Bohemian farms, which was replaced in time with a book called *Pond Fish Culture*, or *Fish Farming*. This little volume first appeared in

1869, but was written by a well-known encyclopedist called Špatný, rather than by a fish farmer. Later, in 1884, *On Nutrition of the Carp and its Fishpond Associates* came onto the shelves. This detailed volume was written by the famous Josef Susta, the director of the large Trebon Fish Farm. Susta was recognized as the principal figure in the national effort to revitalize the industry, particularly through intensification and increased pond productivity. To these ends, he was assisted by Antonin Fric and Josef Kafka, two hydrofisheries biologists from the National Museum of Prague who established field stations for testing water and soil chemistry around the different regions where there were ponds.

Far away in Spain, Mariano de la Paz Graells published his *Manual practico de Piscicultura* in 1864, and two years later, built a small laboratory near Segovia for the culture of trout to stock local lakes and streams. His work was so successful that the Cuerpo de Ingenieros de Montes, a government body, was given added responsibility for fish breeding and stocking all waters within their jurisdiction. In France, there appeared several articles by Monsieur G. Millet, mostly on the theme of repopulation of both inland and marine waters. In the British Isles, there were two separate but similar works published by Edmund and Thomas Ashworth, and Robert Ramsbottom on the propagation of salmon. William Brown published a book about salmon-work on the River Tay at Stormontfield, Scotland, and Francis Francis wrote another on fish culture. There were also in print translations into English of some of the early manuscripts by Gottlieb Boccius, Géhin and Remy, and Coste.

Unquestionably, men like Susta, Coste, Holmberg, and Buckland were among the "fathers" of fish and shellfish culture in the Old World, not only because of their first-hand knowledge of the subject, but also because of their public endeavors to make propagation a useful option for fisheries management. Although there had been significant advances in science, particularly in natural history, for a century or more, Buckland might also be called the father of fisheries science. It was due to his persistent quest for factual information and evidence of the state of indigenous populations of fish, crustaceans, and mollusks (and even the Arctic seals of Jan Mayen) in the North Sea and around the British Isles that the need for research to direct fisheries policies was first established. His memory is aptly preserved through the Buckland Lectureship on Economic Fish Culture. In his will, Buckland gave the nation a sum of money to establish the Professorship of Economic Fish Culture. The Buckland Foundation, which now administers the trust, annually appoints the Buckland Professor, who has the responsibility of delivering an appropriate lecture on fisheries management and science.

Scientific research and fish production techniques were not only confined to the new focus on enhancement of the natural populations of fish. There were also new benefits for fish farming. The traditional warm-water fishes were still the principal farmed fish of Eastern Europe. In 1870, progress was greatly hastened by Thomas Dublisch, working at Cieszyn in southern Poland. He developed a management system for transferring carp fry from the breeding ponds to small fry-rearing ponds, which were suitably prepared in advance, and thence to larger nursery ponds. The word spread quickly among fish culturists as a result of events such as the Vienna Exposition in 1873. Dublisch's techniques greatly increased the survival of warm-water fish in Europe and significantly increased the overall yields of the fish farms.

However, an even more important change in European fish farming was in the offing. In 1879, the U.S. Fish Commission shipped rainbow trout eggs to the Trocadéro Aquarium in Paris, where they were raised successfully. The event immediately attracted attention. Further shipments were sent to Germany in 1882 and 1884, from where the eggs of subsequent generations were successfully introduced and established in Austria, Bulgaria, Czechoslovakia, Denmark, Hungary, the Netherlands, and Sweden. The National Fish Culture association in England received the first of regular shipments from the United States in 1884, and soon the resulting egg crops were distributed throughout all the British Isles. Almost everyone had the same goal—namely, to stock every suitable river and lake to let the anglers of Europe sport with this fighting game fish. One exception was Denmark. The Danes had a number of hatcheries for indigenous brown trout, which they had been raising for some time. These hatcheries were quickly adapted for rainbow trout in the early 1890s. But instead of following the common practice of releasing the young fish to enhance the local watersheds, the Danes continued to keep and feed the fish in captivity. These grow-out hatcheries became the first land-based fish farming enterprises in Europe. Other countries quickly followed suit, but many of their hatcheries did not have the additional water capacity for farm production. In Spain, a private hatchery for trout culture that had been built in 1886 at the Monasterio de Piedra near Zaragoza was later converted to farm rainbow trout, but the Danes already had the lead they needed to be able to dominate the farm trout industry in Europe for the next hundred years.

5.3   Advances in the New World

At the same time as the early beginnings of modern fish culture were developing in parallel all over Western and Eastern Europe, inquiring minds in the New World were carrying out the same simple breeding and rearing experiments with equal success. In Ohio, a surgeon and amateur naturalist called Theodatus Garlick artificially bred brook trout from eggs that he had brought back from Canada in 1853. His hatchery-reared fish, raised on the estate of his good friend Professor H.A. Ackley, soon became the excitement of the annual agricultural state fairs, a characteristic of American rural life that persists to the present day.

The successful initiative of Garlick and Ackley was but one of many that were part of an increasing scientific interest in the natural history of fish. Within the next twenty years, the eggs of a number of common freshwater fish had been reared in primitive laboratory conditions and the young fry put back into their natural waters before the fish died in captivity. The majority of these species were the salmonids, all popular among sport fishermen then as now. These common fish, such as the brook trout, rainbow trout, brown trout, shad, and salmon, were readily caught; their eggs were large and visible; and the emergent fry had a plentiful supply of food from the yolk, which avoided the complication of feeding.

Many of these early efforts were supported by landowners and gentlemen sport fishermen who immediately recognized the potential of hatching and re-stocking to increase the value of their properties and to improve the sport fishing for the benefit of themselves and their friends. With independent financial resources to build their own hatcheries, these individuals started many small programs of enhancement. Word of their activities was also spread far afield, and transplantation of stocks was made not only to friends at home, but also to friends and fellow countrymen who had emigrated to distant countries.

The first true fish culturist in North America was probably Richard Nettle, a civil servant working in Quebec and a personal friend of the governor. Thus, it was easy for him to get permission to build a small hatchery facility in 1856 on the St. Charles River, near Quebec, first to raise trout, followed by Atlantic salmon. He was successful in raising fingerlings, but his work was abandoned a few years later, because it was thought to be impractical. For his pioneering work, Nettle was made superintendent of fisheries of lower Canada in 1857, and he continued to influence fisheries development in the country.

In 1868, successful rearing of the Atlantic salmon was reported again on the Miramichi River. However, by that time, the first hatchery for Atlantic salmon was already under operation in Ontario. Samuel Wilmot built the hatchery by himself in 1865 on his own property adjacent to the shores of Lake Ontario, and he made traps to catch the fish on their returning migrations. Wilmot had been experimenting in his basement for some years and had been successful in raising fish from eggs incubated in trays and later releasing them as parr. He obtained his first smolts in 1868 and subsequently released them into the creek. In 1867, with money from the government, the Wilmot's Creek (Newcastle) hatchery was expanded to produce in time about one million salmon each year. This was in all probability the first actual ranching of salmon raised and released from a hatchery. Samuel Wilmot himself went on to become an influential man in the development of fish culture in the new Dominion of Canada and also in the management of inland fisheries.

Wilmot's work on Lake Ontario regrettably came to nothing in the end, because the gradual build-up of pollution in the lake terminated this first public venture. However, with the needs and benefits clearly spelled out, there was no turning back. By the end of the century, the Maritime Provinces of Canada had several hatcheries, and the movement spread to the west with the construction of a hatchery for Pacific salmon on Harrison Lake in British Columbia in 1884.

Across the border in the United States, the concern for the inland fisheries and interest in fish culture were so strong that in 1871, the government created the U.S. Fish Commission. Efforts to sustain the Atlantic salmon closely followed those in Canada, and the first hatchery was built in Maine in 1870. But the smaller shad was the basis of large commercial fisheries on the East Coast and therefore received considerable attention in its own right. For almost a decade, fish culturists had been taking and transplanting eggs of the shad after fertilization. Of course, there were many failures. Eggs were introduced without success

(a)

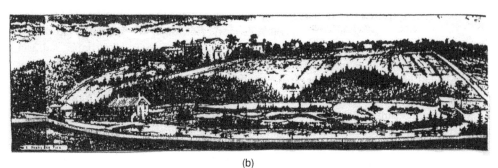

(b)

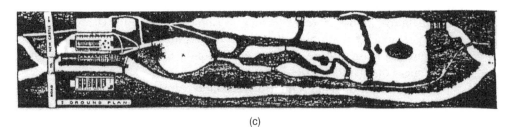

(c)

Figure 5.3 Canada, 1877; Dominion Salmon Culture Station at Newcastle Creek, Ontario: (a) exterior of the hatchery building; (b) bird's-eye view of the station; (c) ground plan. (Courtesy of Stephen Crawford, University of Guelph, Ontario, Canada.)

into southern rivers emptying into the Gulf of Mexico in 1848, and likewise, the many attempts to introduce them to the Colorado River and Great Salt Lake between 1873 and 1892 were failures.

The first successful experiments for the propagation of the popular shad in the United States took place in 1867 on the Connecticut River, which opened the way for one of the most important transplants on record. The man behind this experiment was Seth Green, a fish culturist from Rochester in New York. With the aid of simple milk cans, he shipped twelve thousand shad fry from the Hudson in eight days and released them into the Sacramento River at Teham, in June 1871. The transplant was so successful that five years later, the first shad were caught as far north as the Columbia River, having rapidly moved up the Pacific coast and spread to create an important regional fishery.

Another pioneer culturist of the time was Livingston Stone. Because of his experience, Stone was appointed a deputy by the head of the newly created U.S. Fish Commission, Spencer Fullerton Baird, and asked to establish an egg-collecting station specifically with the intention of introducing Pacific salmon into the East Coast rivers to replace the dwindling resources of Atlantic salmon. As director of the National Museum and secretary of the Smithsonian Institution, Baird was a powerful political force as well as an enthusiastic and capable naturalist. He had written the bill creating the U.S. Fish Commission and helped push it through the United States Congress. As commissioner, he appointed several of the leading fish culturists of the day into staff positions to fulfill one of its primary goals: namely, to increase the availability of food fish. Within this broad mandate, Stone located the U.S. Fish Commission's first Pacific salmon egg-taking station and subsequently a hatchery on the McCloud River in northern California in 1872. From the station, later named the Baird hatchery, he shipped fertilized salmon eggs around the state, into neighboring Utah, and back across the continent to New England. Eggs from McCloud eventually would cross the Pacific Ocean to New Zealand. Stone was also responsible for other transcontinental introductions, including that of catfish and smallmouth bass into California in 1874, and striped bass in 1879.

On December 20, 1870, at the headquarters of the Poultry Society in New York, Baird and Stone, together with other fisheries biologists and naturalists of the day, formed a modest professional club called The American Fish Culturists' Association. The Association met regularly in different locations to discuss issues and projects, which Baird then funded through the U.S. Fish Commission. In 1878, the name was changed to the American Fish Culture Association. This was still not satisfactory to all of its members, many of whom were not culturists, but rather, wildlife naturalists and whose interest extended beyond the food fish species. There was much discussion at the meetings to change the name again, with strong support to maintain the prime interest of fish culture. However, faced with the choice of American Fisheries Association and American Fisheries Society, the president pushed through the latter in committee in 1884, and Stone became the first secretary.

It was not long before the attention of the U.S. Fish Commission turned to the rescue of the world's greatest salmon river, the Columbia, and the many tributaries that formed the vast Columbia Basin. Stone built the first true hatchery for Pacific salmon on the Clackamas River in Oregon in 1877. By the end of the century, the state of Washington alone had fourteen salmon hatcheries producing over fifty-eight million fry.

Common carp was first introduced privately from Germany to California in 1872 by Julius Poppe, an immigrant living in Sonoma County. With the

Figure 5.4 United States, McCloud Hatchery, California, later renamed the Baird Hatchery (now beneath Lake Shasta).

help of Captain John Harlow and the U.S. Fish Commission, the common carp was repeatedly dispersed widely through the western states between 1876 and 1880. Dispersion was also helped by a flood that inundated Harlow's ponds at Troutdale in Oregon and liberated his stock into the interconnecting waters of the Columbia Basin. Here the carp flourished. The 1899 Willamette River catch in Oregon was so good that many fish were iced down and exported back to Germany for market.

From the egg-taking stations and hatcheries being built all over the country, fertilized fish eggs were shipped liberally from place to place. There were no limits to what was tried. Rainbow trout and brook trout were transplanted from the United States across the Atlantic Ocean to the trout hatcheries of Europe. Eggs were also carried across the Pacific Ocean by fast clipper, and New Zealand received its first shipment of salmon eggs in 1868. This was later followed by the first transplants to Japan of rainbow trout and brown trout eggs in 1877.

Neither the pursuit of science and sport fishing nor the longing for fish from the Old World by homesick expatriates was the sole justification for an interest in fish culture. In North America, the onset of logging and degradation of the natural spawning grounds of many of these popular species, together with indiscriminate and heavy fishing to meet the needs of a population rapidly increasing with immigrants, were causing substantial reductions in once-productive native fisheries. The damming of rivers and waterways for agricultural irrigation and navigation also heralded the start of man's ever-increasing competition with fish for the use of water. The artificial culture of fish and the enhancement of fisheries were obvious and easy solutions in the emerging armory of management practices to compensate for the damage.

Bibliography

Ashworth, E. and Ashworth, T. (1853) *A Treatise on the Propagation of Salmon and Other Fish*. Simpkin and Marshall, London. Available at: http://books.google.com/books?id=vyEoAAAAYAAJ&dq=Treatise%20on%20the%20Propagation%20of%20Salmon%20and%20other%20Fish%2C%20Ashworth&pg=PP1#v=onepage&q&f=false (last accessed on August 6, 2010).

Boccius, G. (1848) *Fish in Rivers and Streams: a Treatise on the Management of Fish in Fresh Waters, by Artificial Spawning, Breeding and Rearing; Showing also the Cause of the Depletion of all Rivers and Streams*. J. Van Voorst, London.

Brown, W. (1862) *Natural History of the Salmon, as Ascertained at Stormontfield*. William Brown, Glasgow.

Buckland, F. (1863) *Fish Hatching*. Tinsley Brothers, London.

Chi Han (300) [uses of seaweed]. Cited in: Chi Han (ca. 300) *Nan-Fang Ts'ao Mu* (Trees and Plants of the South) [in Chinese].

Confucius (ca. 500 BC) [poetry]. Cited in: Schwimmer, M. and Schwimmer, D. (1955). *The Role of Algae and Plankton in Medicine*. Grune & Stratton, New York.

Coste, M. (1853) *Instructions Pratiques sur la Pisciculture (Practical Instruction on Fish Culture)*. Académie des sciences, Paris [in French].

Coste, M. (1861) *Voyage d'Exploration sur le Littoral de la France et de l'Italie (Voyage of Exploration on the Shore of France and Italy)*. Imprimerie impériale, Paris, France [in French].

Darwin, C. (1859) *The Origin of Species by Means of Natural Selection, or the Preservation of Favoured Races in the Struggle for Life*. John Murray, London.

de la Paz Graells, M. (1864) *Manual practico de Piscicultura (Practical Manual of Fish Culture)*. Spain [in Spanish]. Available at: http://books.google.com/books?id=VU4rAAAAYAAJ&ots=KA71Edui7v&dq=Manual%20practico%20de%20Piscicultura%20by%20Mariano%20de%20la%20Paz%20Graells&pg=PR3#v=onepage&q&f=false (last accessed on August 6, 2010).

de Quatrefages, A. (1848) Des fécondations artificielles appliquées a l'élève du Poisson (Artificial fertilization applied to the breeding of fish). *Comptes rendus de l'Académie des sciences 1848*, Paris [in French].

du Monceau, D. (1763–1782) *Traité Général des Pêches et Histoire des Poissons (General Treatise on Fishing and the History of Fishes)*. Académie des sciences, Paris.

Francis, F. (1883) *The Practical Management of Fisheries: a Book for Proprietors and Keepers*. London. Available at: http://books.google.com/books?id=im8oAAAAYAAJ

&dq=The%20Practical%20Management%20of%20Fisheries%3A%20a%20Book%20for%20Proprietors%20and%20Keepers.&pg=PP2#v=onepage&q&f=false (last accessed on August 6, 2010).

Géhin, A. and Remy, J. (1851) Fécondation artificielle des poissons (Artificial fertilization of fishes). *Journal des Traveaux de l'Académie Nationale* [in French].

Horák, W. (1869) *Die Teichwirthschaft mit besonderer Rücksicht auf das südliche Böhmen (The Pond Industry, with Special Consideration of Southern Bohemia).* Salvesche Universitätsbuchhandlung, Prague [in German].

Jacobi, L. (1763) (Letter essay) *Hanover Magazine* [in German]. Also in: Goldstein, Count de (1770) *Abridgement of the Memoirs of the Berlin Academy* [essay translated to Latin]; du Monceau, D. (1763–1782) *Traité Général des Pêches et Histoire des Poissons* (General Treatise on Fishing and the History of Fishes), Part II, p. 209. Académie des sciences, Paris [translated from Latin to French].

Knight, W. (2007) Samuel Wilmot, fish culture, and recreational fisheries in late 19th century Ontario. *Scientia Canadensis: revue canadienne d'histoire des sciences, des techniques et de la médecine (Scientia Canadensis: Canadian Journal of the History of Science, Technology and Medicine).* 30 (1), 75–90. Original source: Wilmot, S. (1878) *Report of Fish-Breeding in the Dominion of Canada, 1877.* Maclean, Roger & Co., Ottawa.

Leclerc, G.L., Comte de Buffon (1749–1788) *Histoire Naturelle, Générale et Particulière (Natural History, General and Specific),* 35 vols. Académie des sciences, Paris [in French].

Linnaeus (von Linne, C.) (1735) *Systema Naturae (System of Nature),* 1st edn. Netherlands. Additional information available at: http://www.linnaeus.uu.se/online/animal/1_1.html (last accessed on August 5, 2010).

Mendel, G. (1866) Experiments in plant hybridization. Proceedings of the Natural History Society of Brünn. Brünn, Moravia [now Brno, Czech Republic].

Millet, G. (1854) Essais de pisciculture: fecondations naturelles et artificielles des oeufs de poisons (Experiments in fish culture: natural and artificial fertilization of fish eggs). *L'Institut* 22, 257–259 [in French].

Ramsbottom, R. (1854) *The Salmon and Its Artificial Propagation,* 8 vols. London.

Shen Nung (ca. 3000 BC) [prescribed seaweed as medicine]. Cited in: Porterfield, W.M., Jr. (1922) *References to the Algae in the Chinese Classics.* Torrey Botanical Society, Lawrence, Kansas.

Spallanzi, L. (1768) [Artificial fertilization of fish eggs] *Essay on Animal Reproductions.* [translated by M. Maty]. T. Beckett, London.

Špatný, F. (1869–1874) Rybníkárství (Fish Farming). *Živa, the Scientific Collection.* Edition of The Museum of the Czech Kingdom, Prague [in Czech].

Susta, J. (1884) *Nutrition of the Carp and its Fishpond Associates.* Ceskoslovenska Akademie Zemedelska, Prague [in Czech].

von Schönfeld, J.G. (1773) *Die Landwirthschaft und deren Verbesserung nach eigenen Erfahrungen Beschrieben (Agriculture and its Improvement, Described Based on Personal Experience).* Breitkopf, Leipzig.

Yarrell, W. (1836) *History of the British Fishes.* John Van Voorst, London.

Chapter 6

Farming the Sea (1880–1920)

Abstract

The International Fishery Exhibition of 1883 in London, featured marine fish farming technology and the need to replenish marine fisheries. Participants were inspired to start fish and shellfish facilities in their home countries. The earliest marine laboratories were built in Naples and Monaco, funded characteristically by scientists and amateurs. Creation of the U.S. Fish Commission recognized government's role in fish culture. Government funds constructed marine fish hatcheries at Woods Hole and Gloucester Harbor. The Marine Biological Association of England was founded in 1884. British facilities were developed by government, scientists, and fish culturists. Cooperative marine study areas were identified when international conferences met in Stockholm (1899) and Norway (1901). Marine hatcheries created new demands for engineering: winterized outdoor ponds, specialized egg incubators, and mechanical saltwater systems. Marine fish culture was promoted to replenish natural fisheries with fertile eggs and hatchery-raised larvae. By 1914, most facilities thus built had failed.

6.1   The International Fishery Exhibition of London

Although Frank Buckland began his interest in the culture of the common freshwater fishes, particularly the salmonids, to restock the depleted natural fisheries of European waters, he turned his attention to farming the sea in his later years. In this, he was not without opponents. Several influential scientists, including the national figure Thomas Huxley, stated that farming the sea was useless, because the sea fisheries were inexhaustible. However, this was too much of a

The History of Aquaculture. By C. E. Nash. Published 2011 by Blackwell Publishing Ltd.

generalization, as noted by one of Buckland's proponents, Ernest Holt. Although it was true that the great fisheries for cod, herring, and pilchard were seemingly beyond the influence of man either by overfishing or enhancing, there were several smaller but important fisheries in the North Sea, such as the flatfish, that could be affected. Holt went on in his lifetime to demonstrate with well-documented statistics the effects of overfishing on the plaice fishery, and in 1897, he proposed solutions for management. Of the four recommendations he made, one was for artificial propagation.

Buckland was but one of many individuals who had turned their attention to fish culture after 1850 with obvious success. The great International Fishery Exhibition of 1883 in London, which had been the forum at which Huxley had made his statements against the need for marine fish farming, devoted considerable space and attention to the new technology. The exhibition ran for almost six months, from mid-May until the end of October, and was attended by scientists and fisheries delegates from thirty-one nations and colonies. There were many demonstration stands from the United States, and several American professors set up models of hatcheries they used to produce and release the juveniles of shad or whitefish (called whiting in Europe).

The International Fishery Exhibition of 1883 had far-reaching consequences. Many of the visitors were fascinated and intrigued by the prospects of fish culture and returned to their own countries to establish similar programs. One such man was L.F. Ayson, the principal fisheries scientist in New Zealand, who took back with him on his ship several million salmon eggs in a refrigerator. Another was Lachlan Maclean, who introduced the culture of trout to the Cape Colony in South Africa. After many failures with the shipment of eggs from Scotland, and several other disasters in a series of temporary receiving places between 1884 and 1894, finally, in 1895, the first brown trout and Loch Leven trout parents were successfully stripped of their eggs and milt, and the first "native" fish were reared. By the turn of the century, a proper hatchery had been built and a program established to stock most of the suitable rivers of the Cape Colony and many of its neighboring provinces. Although repeated efforts were made with Atlantic salmon, these were never successful.

The story of the successful introduction and acclimatization of trout into South Africa, achieved by the persistence and endurance typical of the colonial expatriates, was being repeated in Australia, Tasmania, New Zealand, and in several Latin American countries. Furthermore, their efforts also extended to shellfish, particularly to oysters for the production of pearls as well as food. Many of their enterprises were singularly successful.

6.2   Marine fish culture and the coastal hatcheries

With the rapidly spreading interest in fish propagation and fisheries research throughout the last half of the nineteenth century, it was quite apparent that there was a need for specially selected and constructed places where the work was to be done. Temporary accommodation in fish processing plants and

Figure 6.1 South Africa, 1880; surviving trout hatchery at Stellenbosch.

waterside mills proved to be totally inadequate. Propagation needed hatcheries, and fisheries science needed institutes and laboratories. Consequently, one of the final contributions of the golden years was the foundation of many of today's most famous hatcheries, and marine biological and fisheries stations.

France was one of the first countries to construct facilities directly in support of marine fish and shellfish culture. Following the successful work of Professor Coste to salvage the French oyster fisheries, coastal centers for production and some applied research were built at Concarneau in 1859 and at Arcachon in 1863.

Characteristic of this early period, long before the advent of government help, scientists and amateurs invested their own personal wealth in the construction of new facilities. Two new stations were built in the Mediterranean. In some respects, it was surprising that they were the first, or at least among the first true marine science research laboratories, because the Mediterranean was not a focal point of fisheries concern at that time. It was, however, a region that had a singular geographic identity, and there was strong scientific interest in its unique oceanography and natural history. The earliest marine laboratory was probably the Stazione Zoologica in Naples, Italy, founded in 1872. It was funded principally by the German zoologist, Anton Dohrn, who persuaded the Italian government to contribute some of the prime coastal land for the laboratory. Although not actively involved in fish culture, the laboratory had a large and famous aquarium. This was soon followed by a second laboratory in Monaco. This facility had the personal interest and financial support of Prince Albert I, who was himself a scientist of no mean repute.

Government support was more forthcoming in the United States than in Europe. Funds were provided to construct a small but fully operational marine fish hatchery in 1855. Wood's Holl (later Woods Hole), a sheltered location on the Massachusetts coast, was chosen for the hatchery. The water conditions of the site were good, and there was already a small nucleus of scientists carrying out research on marine fish. The early results were so encouraging that the government funded a second hatchery nearby at Gloucester Harbor. This was the first facility for work on the cod, and propagation was successfully achieved in 1878 using the techniques first developed in 1866 by Gunnar Sars, a professor at the university in Christiana, for raising lobsters. This was rapidly followed with successful propagation and release of other gadoids and even herring, under conditions that nowadays would be described as relatively poor. Gloucester Harbor hatchery was subsequently enlarged twice to cope with the great diversity of species it was producing.

With this spectacular start to marine fish culture in the United States, immediately on top of a successful freshwater fish culture program with the transplantation of the shad from the East to the West Coast, the benefits of fish culture to commercial fisheries were readily accepted without further question. Upon the creation of the U.S. Fish Commission in 1871, fish culture was recognized as an official responsibility of government in the United States, with its own department. Immediately upon his appointment as the first commissioner, Spencer Baird made a survey of all the important commercial fisheries on the East Coast and reported that marine fish culture could enhance most of these diminishing stocks.

In Europe, immediately following the International Fishery Exhibition, the Marine Biological Association of England was officially founded in 1884 by the Royal Society of London. The first president was the famous Thomas Huxley. The Royal Society, in its first resolution, emphasized the necessity for the establishment of one or more laboratories on the coast of Britain, where "accurate researches may be carried on, leading to the improvement of zoological and botanical science and to an increase in our knowledge as regards the food, life, conditions, and vegetable resources of the sea in general." The first marine biological laboratory was opened at Plymouth in 1888, with three professional staff. In time, it would be complete with its own saltwater aquarium system, library, and research vessels.

The marine laboratory at Plymouth was quickly followed by another that was constructed in the harbor at Lowestoft, in Suffolk. This government-funded laboratory was directed by the naturalist Walter Garstang. From this post, he eventually undertook some fish tagging and transplantation experiments with young fish, particularly the flatfishes, in the North Sea. Two more marine laboratories were established with the support of the Lancashire and Western Fisheries Committee for work relevant to the fisheries of the Irish Sea and the west coast of southern Scotland. One was at Port Erin in the Isle of Man, and the other on Piel Island near Barrow-in-Furness. Both of these laboratories were under the direction of Professor William Herdman, who began substantial propagation

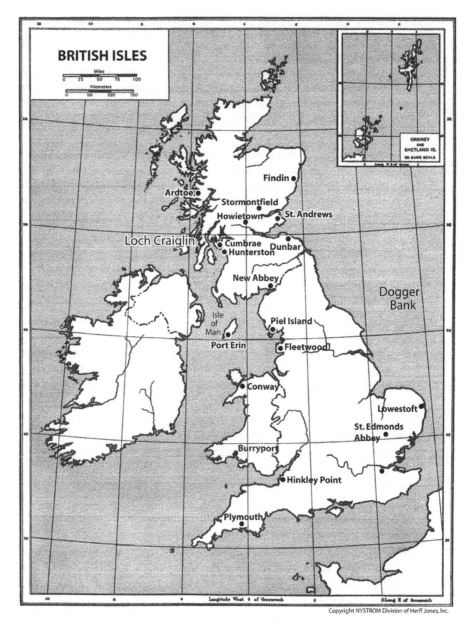

Figure 6.2 British Isles (adapted from basic map, copyright NYSTROM Division of Herff Jones, Inc.).

and release work not only with the flatfishes, plaice, and flounder, but also with cod and haddock.

The first marine fish hatchery in the British Isles was built at Dunbar near Edinburgh in 1894. It was built by J. Cossor Ewart, who had made a prolonged tour of laboratories and hatcheries in the United States and Canada immediately after seeing the displays at the 1883 International Exhibition. The purpose of the Dunbar hatchery, he said, was "to raise and release young marine fish to directly add to the fish supply," and he began with the propagation of the plaice to enhance the coastal fisheries.

Ewart's work at the Dunbar hatchery in Scotland was helped by the presence of Harald Dannevig, who with his father Gunnar Dannevig had built and operated the first commercial hatchery for cod at Arendal on the southern tip of Norway in 1882. The elder Dannevig had also visited the United States and Canada in 1883 after the Fishery Exhibition and returned to make many improvements to his hatchery at Flødevigen, where he was raising fish for release into the fjords around the Skagerrak. Many Norwegians rapidly became interested and skilled in fish culture, and one expatriate, Adolph Nielsen, operated the first marine fish hatchery in Newfoundland for the Canadian Department of Fisheries.

The pristine beaches and rocky coastlines around Scotland encouraged many scientists to turn their attention to the expanding fields of marine science, and there was a clear need for coastal laboratories where they could work. In 1885, through the enterprise of Sir John Murray, an aptly named vessel called "the Ark" was hauled close to the beach on the island of Great Cumbrae in the Firth of Clyde to become a floating laboratory. It proved to be such a considerable attraction to many of the leading scientists of the time that the burghers of Millport persuaded the local entrepreneurs to build a completely new research facility and provide it with a marine aquarium for the holiday-making public. The Millport marine station was opened in 1897 and became an immediate focal point. Within five years, the scientists working there formed the Marine Biological Association of the West of Scotland, which in a matter of a few more years would become the Scottish Marine Biological Association.

For the burgeoning number of marine scientists on the east coast of Scotland, a hospital building at St. Andrews was converted into a temporary laboratory accommodation by Charles Henry Gatty, an enthusiastic amateur who lived in southern England. Just like Millport, the facility would go on to become permanent. Gatty and his friend Lord Reay soon raised enough money to build a new laboratory and public aquarium. The facility was opened by Reay in 1896 and became known as the Gatty Marine Laboratory. After this promising start, the laboratory had a checkered career following a fire, believed to have been started by local suffragettes, until it found shelter with its close neighbor, St. Andrews University.

The Ark at Millport and the old fever hospital at St. Andrew's were just two of several premises that were temporarily occupied by the scientists and fish culturists around the coast of Scotland until 1899. It was in that year that the Fishery Board for Scotland, an august body that had been commissioned in 1882 to identify a site suitable for the country's principal marine research center, settled on Aberdeen.

Despite the numbers of marine science laboratories that were springing up on both sides of the Atlantic and the many directions for research in response to the excitement of the times, by the start of the twentieth century, the marine and fisheries science carried out at the majority of these coastal laboratories was very similar. This was because in 1899, the Swedish government convened an international conference in Stockholm to which most of the leading scientists of the age were invited. The first conference was attended in force by twenty-six delegates from Denmark, Germany, Great Britain, Holland, Norway, Russia, and Sweden. For some reason, Belgium and France did not send any representatives. At the invitation of the newly independent country of Norway, a second conference followed in 1901, with the addition of delegates from Belgium and Finland, but still not France. This illustrious gathering at Christiana (soon to change its name to Oslo) established the common methods for gathering information, and identified specific cooperative programs for biological and hydrographic studies. Among the former were investigations on the life histories of the important fish species, their behavior and migrations, as well as the effects of fishing and fishing gear on populations. Also there were investigations on the marine plankton. The Christiana Program, as it became called, proved to be extremely important in coordinating some of the first fisheries and marine research over an extensive area of the North Sea, the Northern Atlantic Ocean, and the Baltic Sea.

6.3   Advances in fish rearing techniques

The construction of the first commercial fish hatcheries, marine research laboratories, and aquariums, where live creatures were displayed for the public, brought demands for structures and engineering that had not been previously envisaged. Moreover, unlike freshwater facilities that were always sited to obtain water through gravity flow, the coastal marine buildings required mechanical water systems, which had to be strong and reliable. They had to be engineered to deal with tides that might differ by four or five meters in height in a matter of hours and to draw from certain depths the best quality of sea water to create safe environments for securely handling the live animals and plants and for providing the proper conditions for propagation.

Because of the need to hold large broodstock fish in captivity, all the early marine laboratories and fish hatcheries were provided with large outdoor ponds or tidal basins built solidly for protection against heavy winter weather. At Woods Hole in Massachusetts, the scientists working with adult cod once tried to reduce the cost of basins by constructing a number of net-pens in water of ten fathoms and more, but the mortality was so high that the idea was abandoned. All the facilities were engineered with at least two large mechanical pumps to guarantee the provision of a high rate of water exchange almost continuously through each pond and any nearby aquarium.

Most of the ponds and basins of the early hatcheries were modeled after those constructed for adult cod by the Dannevigs at their Flødevigen hatchery in Norway. At first, these outdoor containers were used both for holding broodstock, as well as for spawning and egg incubation. However, it became clear that better control was required for both incubation and subsequent larval rearing. Because most of the eggs of the marine fish under study were pelagic and easily visible, and were individual rather than agglutinated and clumped, it was a simple step for the early pioneers to collect the eggs from the surface of the tanks with fine mesh nets and to incubate them indoors.

The contraptions that the early culturists constructed for egg incubation were very individualistic at first. Each had features that the designer thought important for increasing survival. The need for water of high quality was already well known; several of the first hatcheries had already been relocated to take advantage of better water conditions, but other requirements had not been fully determined. For example, at Woods Hole, Captain Chester used glass jars for the incubation of eggs, and he simulated tidal conditions by siphoning water through a large wooden water-bath in which the inverted jars were placed. The eggs were retained in the jars with a fine cloth. The high rate of water exchange and the movement of the eggs produced a high survival, and "Chester" jars were widely copied.

The tidal egg-hatching box was also popular. This was developed by Marshall MacDonald, also at Woods Hole, who had succeeded as head of the U.S. Fish Commission on the death of Baird in 1887. Because each Chester jar could only hold about 200 thousand eggs, many jars were required to incubate the numbers of eggs that were being produced by the large females of the main fish species under culture. There was clearly a need to enlarge the size of the incubators without lowering survival. The MacDonald tidal box, which was designed with screened compartments and water chambers, more than doubled the capacity of the Chester jar.

In Norway, the Dannevigs developed a rocking incubator that simulated the gentle water movement in the sea and maintained eggs individually and in suspension. The incubator had a capacity of 600 thousand eggs or more and had a marked effect on survival. Inasmuch as the Dannevigs advised many of the European countries on their propagation hatcheries, their effective incubator was very commonly used in European coastal hatcheries for marine fish. It was also very effective, whereas the MacDonald incubator was adopted by all of the American hatcheries.

Common to all hatcheries on both sides of the Atlantic were the mechanical water systems, all of which quickly became victims of natural marine biofouling. To keep out the mollusks and algae that settled, grew, and eventually blocked the inside of the intake pipes, and to prevent entry of other minute marine creatures that survived the force of the pumps to find a new home in the incubation tanks and aquariums, all the water systems had to be fitted with systems for filtration. Iron grills, sand-filters, and settling tanks soon became a necessary but costly part of the intake system for every newly built hatchery, and finer filters were needed for the delivery system to the rearing tanks. And this was not all. The early mechanical seawater systems were made of cast-iron pipe, which although substantial and strong, soon corroded and colored the water with rust. Even the valves and tank fittings, which were frequently made of costly brass or stainless steel, were quickly covered with a fine film of corrosive salts that slowly ate them away. Worse still, these hydrated metallic salts also dripped or leached directly into the hatchery tanks, causing problems of metal toxicity, which were not discovered and overcome for another half century.

Consequently, the early years of hatchery technology were years of much experimentation in both equipment design and materials, as well as in biology and culture. It was soon obvious that marine fish hatcheries were not simple facilities to design and put together, nor were they cheap to operate and maintain. Fortunately, many governments by then had established fisheries policies, and fish propagation was an active component of fisheries management. As a result, at the turn of the century, almost all the once-private hatcheries were taken over by governments, and governments provided the funds for all new hatcheries.

6.4 Propagation and larval releases

Marine fish culture was promoted by the pioneers to replenish the natural fisheries with fertile eggs and young larvae, which were otherwise lost by overfishing of mature fish or by the capture of undersized fish, due to the lack of net regulations. The premise of these early fisheries scientists was that the annual brood strength of the fishery was directly proportionate to the number of eggs released by the adult population. A number of solutions were therefore offered, but only for the purpose of compensating the losses, first with eggs and later with hatched larvae.

Transplantation of marine fish, particularly of young plaice, was first started in Denmark as a private initiative by Justitsraad Hansen of Thisted. In 1892, he formed the Association of Thisted Fishermen to help carry out the work of stocking the local waters. The benefit of transplanting young fish to new rich feeding grounds was immediately apparent in their enormous growth rates. The Danish government stepped in, and the scope of the work was extended to the larger Limfjorden, the vast inland sea that almost isolates the northern tip of the country. Meanwhile, Gunnar Dannevig had been transplanting hatchery-bred cod in Norway, and consequently, he too moved across the Skagerrak into the Limfjorden. Between the efforts of the Danes and the Norwegians, many millions of wild-caught and hatchery-raised fish were transplanted almost without a break until the last years of the First World War.

Their work sparked interest among fisheries and marine scientists in England, Sweden, and Germany. In England, Walter Garstang went further, and in 1904, he began transplanting small wild plaice caught in the coastal estuaries to the Dogger Bank at the northerly end of the English Channel. All these transplants, many with samples of marked fish, were accompanied by supporting research that showed that the fish grew well and often better than control stocks. For example, in Limfjorden, the two-year-old plaice typically increased their length by 50% in a six-month growing season, and on the Dogger Bank, even larger fish transplanted from the surrounding coasts gained some two to three times more than their normal seasonal growth increment.

Walter Garstang was a marine scientist as famous for his wonderful poetic verses about the creatures of the plankton as much as for his science and zeal for restocking and transplantation. Keenly aware of the importance of the density of planktonic food organisms for the survival and growth of young fish, he proposed that all fishing vessels should be equipped with buckets and tanks so that trained fishermen might strip eggs and milt from suitable fish, fertilize the eggs, and then return them to the areas of the sea rich in food. Garstang's suggestion was widely supported, and he himself took many voyages on commercial vessels to demonstrate the technique.

However, most of the new group of fisheries scientists in Europe, particularly the aging Buckland and Ernest Holt, favored artificial propagation rather than transplantation. Mature fish were caught and allowed to breed naturally in captivity or stripped of their gametes; fertilized eggs were collected and incubated, and the larvae retained for a short period to complete metamorphosis before release into the sea. With no more than these early life stages as the targets of production, the marine hatcheries that were rapidly constructed all over the world had no difficulty in realizing some very impressive numbers. With only a modest number of broodstock, significant numbers of fry could be released. For example, three hatcheries on the eastern seaboard of the United States annually accounted for some three billion fry, primarily pollock and flounder. This more than dwarfed the efforts elsewhere, but all were considerable. In Australia, under the expert eye of Harald Dannevig, who had emigrated there in 1902, the hatchery at Gunnamatta Bay in New South Wales was producing 150 million annually, and this was emulated by a new hatchery at Dunedin in New Zealand.

The euphoria of marine fish culture and enhancement did not last. In the end, most of these pioneering enterprises were not successful. There was still a great lack of understanding about the behavior, reproductive biology, nutrition, and diseases of the fish and shellfish that were being propagated. Moreover, the hatcheries were expensive to operate and maintain, and without any strong evidence that there was a positive benefit to the fisheries, many people began to question their cost-effectiveness. By the beginning of the First World War, the majority had closed down or had been converted to laboratories for fundamental research in marine science.

Chapter 7

Fifty Lost Years (1900–1950)

Abstract

Fish farming and shellfish production infrastructure was built between 1900 and 1950, but without real market. Investment was discouraged by political upheaval, world wars, and global depression. Colonial African fish culture allowed European introduction of species for sport fishing and for controlling insect-borne disease. Government fish farms spawned fisheries departments and research stations. Japanese fishermen applied seaweed and shellfish culture. The Meiji government developed experimental hatcheries; almost all eel production was cultured. The revolutionary technique of oyster hanging-culture emerged. Under postwar democratization and land redistribution, fishermen's unions embraced aquaculture. Freshwater fish culture in the Pacific Northwest helped counter impacts to salmon from dams, irrigation, and poor logging and mining practices. The 1934 Fish and Wildlife Conservation Act gave legal authority to protect and compensate for salmon fisheries affected by federal water projects. Other laws strengthened mitigation and restoration. New pelleted feeds benefited all of aquaculture. Brine shrimp nauplii, dry-storable until needed, were identified as live food for marine fish larvae.

7.1 Introduction

In the emergence of modern aquaculture, the first half of the twentieth century can be described with some generosity as a period noted more for building infrastructure than for the achievement of great scientific and technical advances. As the previous century closed on its new discoveries of breeding and propagation,

The History of Aquaculture. By C. E. Nash. Published 2011 by Blackwell Publishing Ltd.

word of the new technology spread rapidly; thanks in no small part to the forums of the international exhibitions in Edinburgh and London.

The importance and potential of these new discoveries were not fully realized. There was still no real need for cultured fisheries products on the large Western markets, and the momentum of discovery alone could not be sustained by the few champions of fisheries management because of the lack of any immediate and obvious success. Furthermore, the investment required for commercialization of this alternative form of fish and shellfish production was continually discouraged by political upheaval in Europe, two world wars, and a decade of global depression.

Nonetheless, between 1900 and 1950, there were many isolated events that usefully consolidated the technology in many countries and had several long-term benefits—some of which would not be realized for another fifty years.

7.2   The influence of the colonial empires

The turn of the century was a time when the influence of the European colonial empires in Africa and Asia was at its peak. Most of the early problems associated with territorial occupation were under control, all infrastructures were well established and operating smoothly, and the administrators were full of confidence. With more and more time to relax, many public servants responsible for these overseas territories were intrigued by the new field of fisheries science and the technology of fish culture that had been capturing the imagination of their kinsmen in Europe. For some colonialists, this was an opportunity to look at the indigenous fisheries of the great African lakes in a new light and to study them scientifically. For others, it was an opportunity to bring the old familiar fish from home to rear and stock in local rivers to add the leisurely pastime of fishing to their other country-club pursuits, such as polo, cricket, tennis, and golf. But that was not all. On a far more serious note, there was also an opportunity to kill a very pestilent nuisance with this same stone.

The early part of the twentieth century was the glorious age of steam ships. Everyone and everything moved by water. Sumptuous liners transported the colonialists and their families to their overseas posts, and tramp steamers carried back the plundered resources to feed the industries at home. Consequently, the majority of these expatriates lived in cities and towns close to water, where the flies and mosquitoes were a constant scourge to their health. In the intense heat and humidity of the monsoon months, the administrators *en masse* moved to upland regions, away from the wet lowland plains, where the risks of malaria, river blindness, and a host of other tropical diseases were at their greatest. Therefore, they thought quite logically, if fish culture technology could be used to introduce European trout and salmon to the cool mountain streams tumbling down from the hills, then it could also be used to introduce insect-eating fish to clear the static, infested waters down on the coast.

Because one of the principal preoccupations of the respective colonial services was with human health, the control of the flies and mosquitoes that transmitted parasitic diseases became a natural responsibility of the governor and his administration. Consequently, the colonial civil servants were co-opted into fulfilling broad government policies to improve public health by introducing mosquito-eating fish into local water bodies, specifically for the control of malaria and other waterborne diseases. Later, the same officials would be called on to help produce fish to supplement the protein requirements of the indigenous people, once protein was no longer adequately supplied through the traditional practices of hunting. Any issue of public health was sufficient justification to have colonial administrations pay first for the construction and operation of fish hatcheries to produce mosquito fish, then for government farms for fish production, and along with these, to provide for regulation of the fisheries and organization of sport fishing.

When fish suddenly took on these new and important roles, it became expedient for the colonial administrators to establish a new department in the government structure devoted entirely to inland fisheries, and then to appoint a hierarchy of fisheries officers. They also built small research centers and government research farms to follow up discoveries that were being made in Europe. These new fisheries departments in territorial African and Asian countries were structured in the same manner as their equivalents in the home countries of Europe. They operated with directors and officers, ranked and promoted exactly as were their counterparts back home. Many no doubt preferred the chance to work with the familiar European species, such as trout and carp, rather than with unfamiliar indigenous charges, such as tilapias and catfish.

The introduction of species exotic to Africa was particularly evident in the colonial countries of Great Britain. The first recorded introductions of temperate-water sport fish, particularly salmonids, to Africa were mostly confined to the central eastern countries that had suitable uplands and mountain streams, and many of which had little indigenous fish life. These areas stretched from Kenya down to South Africa and included countries such as Uganda, Nyasaland (Malawi), Northern Rhodesia (Zambia), Southern Rhodesia (Zimbabwe), and Swaziland.

The first introduction of trout in Kenya, for example, was probably typical of the times. It was made by Major Ewart Gordon in 1905. With the help of his sporting friends, he brought fertilized eggs of brown trout from Loch Leven in Scotland, then incubated and released the fry into the waters of the River Gura. Reportedly, they flourished and became so abundant that they rapidly consumed all the available food. The situation soon got out of hand and could no longer be left to amateur enthusiasts. It became necessary to place the fisheries under the control of the Forestry Department and to make their supervision the responsibility of a game warden. A hatchery was eventually built and the fish distributed among other streams, but it would be another decade before sport fishing in the region was sensibly organized and managed by the Kenya Angling Association. Further introductions into Kenya were much more rigidly controlled.

Introduction and transplanting of species of fish in Africa was not only confined to the salmonids of interest to anglers. Records also show the introduction of many other species, such as the Chinese carps, and the liberal transfer and introduction of many native species, such as the cichlids, to entirely new ecosystems. In Egypt, North Africa, where the British still held tenuous occupation, the Barrage Fish Farm was built in 1929 at El Kanatir el Khairia, not far from Cairo. It received introductions and produced fish seed for stocking the country's large coastal lakes. This was followed in 1931 by the El Mex Fish Farm, constructed alongside one of the principal irrigation pumping station near Alexandria. Both of these farms were part of the country's continuing irrigation and fisheries compensation program in the Nile Delta.

Even more adventurous at that time was the attempt by British and Egyptian scientists to relocate marine species around Egypt. Gray mullets (1928) and Dover sole (1938) were released into the highly saline waters of inland Lake Qarun, located eighty-three kilometers southwest of Cairo. The low-lying Lake Qarun was a major drainage sump for agricultural runoff, and the evaporation and concentration of salts enabled the successful establishment of new but small "marine" fisheries that were to last for the next fifty years. Much later, in 1977, as the salinity increased to fifty parts per thousand and beyond, some marine shrimps were successfully introduced.

For all the early colonialists and settlers in foreign parts, personal health was always the top priority. Expatriates had little in the way of natural immunity or effective medicines to combat an array of tropical diseases, and therefore, a good healthful diet was important. It was fortunate for them that their foreign destinations were primarily those for which acute shortages of protein were the exception rather than the rule. However, this was not the case on the subcontinent of India.

For a large part of the enormous population of India, eating meat was forbidden by religious law. For the rest, meat was a scarce and costly commodity. The possibility of increasing protein availability through increased fish production in the hinterland was an attractive, practical option for the British administrators. Furthermore, subsistence fish farming of the Indian major carps had become something of a tradition by that time, and production included a small number of exotic species. The gourami was first brought to Calcutta from Java in 1841. This introduction was followed by another delivery of gourami, this time to Tamil Nadu from Mauritius in 1865. However, it would take another fifty years before the species became productively established. The gourami was followed by some of the Chinese carps from Southeast Asia and the cichlid species, tilapia, which was supposedly carried in by the trading boats from East Africa. Small quantities of various fish were introduced and reintroduced to India many times over the years. Their culture, together with that of the native Indian carps, was the basis of considerable technical research. But it was not until the 1950s, and 1960s, when the Central Inland Fisheries Research Institute set about making large introduction of the great carps, grass carp, and tilapias to its Cuttack station in Orissa, that production became significant.

India, probably more than any other country, epitomized the penchant of the British administrators for government. Not only was the organization and management of each sector embedded in a central bureaucracy, this order was replicated one step down at a state level. The first government fish farm was established in 1911 in the state of Sunkesula by the Fisheries Department of Madras. It was quickly followed by fisheries departments, each with their stations for administration and research, in Bengal, Punjab, Uttar Pradesh, Baroda, Mysore, and Hyderabad. This created an extensive structure for the management and regulation of fish farming in these respective states for the rapidly growing numbers of farmers and fishponds. By midcentury in West Pakistan, over fifty thousand ponds were recorded in the Punjab alone.

The Indians themselves greatly appreciated the potential of fish culture for meeting their national needs, and because of some control at the state level, they were able to develop ways of doing things not necessarily considered conducive to public health by European standards. In the early 1940s, for example, the Institute of Sciences was sponsoring research and development of fish farming in close association with studies on the hydrobiology of many polluted inland waters, and in 1944, it organized a symposium on the use of sewage for fish culture at Calcutta.

7.3 The impact of Japan in Asia

Japan before the twentieth century, like the rest of its neighbors in Asia, was an agricultural nation made up of small-scale farmers and fishermen, who for almost one thousand years had been held in a strict feudalistic system. With the official ending of seclusion and start of trading concessions with United States and Europe in 1852, and the beginning of modernization under the Meiji government in 1868, some Japanese were fortunate to benefit from the economic and social reforms that took place. Commoners became tightly bound into their own small, independent communities. The majority, however, still worked as peasants for the newly privileged ruling classes, who owned most of the land.

As a mountainous island nation with almost unlimited views of the sea, Japan developed a culture in which the natural beauty of the rocky coasts has forever played an important part. This is borne out by the number of antique scrolls of seascapes that depicted the activities of fishermen and farmers going about their work. Paintings surviving from the Edo period, or the golden age of art at the end of the eighteenth century, showed that organized production of crops in the sea had been taking place for some time. The production of the edible red laver, for example, shown growing on thin wooden poles and brushwood bundles in the bays of the Inland Sea and up the coast as far as Edo (renamed Tokyo by the Emperor Meiji in 1868) was known to have been practiced when the Jesuits arrived in the sixteenth century.

Wooden poles were an effective means of working the shorelines, and the fishermen applied the same simple technique to rear oysters held in net bags. Paintings also showed fishermen harvesting beds of scallops, but it is known that their attempts at management were not effective, because in some years, the harvests were a disaster. This uncertainty did not change, in fact, until the end of the nineteenth century, when the traditional fishing gear was banned, and orderly use of the beds allowed the untouched scallops to mature. Some of the landscape paintings revealed that freshwater fish culture was also well established, particularly for breeding magnificent varieties of golden carps to keep in the ornamental garden ponds of the rich landowners.

In the two hundred years of Japan's enforced seclusion and the emphasis by the government on internal peace and culture, changes to the *status quo* took place very slowly. It was during this period of tranquility, in 1716, that Buheiji Aoto divided a tributary of the Miomote River, which flowed through his estate, into a few narrow spawning channels to enhance the local run of chum salmon. His methods were successful, but he continued to keep the information to himself. It was not until 1808 that a similar conservation program was started again on the same river in Niigata Prefecture, and it would continue for the next hundred years.

For fish culture, the first small chink in the protective armor surrounding Japan occurred in 1876 when word got through that propagation techniques for trout were very successful in other parts of the world, and Akikiyo Sekizawa recommended to the new Meiji government that it develop experimental trout hatcheries similar to those being built in the United States. The request was timely. It was a period when the government was investing in infrastructure to establish a central economic base and to replace the ancient tax system that benefited only the old *bakufu*. The government agreed, and Sekizawa set to work. The first rainbow trout and brown trout eggs arrived in 1877; these were hatched in his own back garden using water drawn from his well. From there, the juveniles went to rearing ponds of the first small wooden trout hatchery, which was built in the foothills around Ohme, close to the Meiji's new capital city of Tokyo. These fish subsequently formed the broodstock of a larger trout hatchery built by the government at Shirako Kita-adachi in Saitama Prefecture in 1880. The progeny from these first trout transplants to Japan were subsequently released into Lake Inawashiro and Lake Chuzenji. In the same year, the first experimental hatchery for the indigenous Pacific salmon was built at Kairakuen, near Sapporo, at a cost of ¥887. It was built on the orders of Governor Kuroda of Hokkaido following the recommendation of Ulysses S. Treat, an American cannery engineer who had been staying in the port town of Hokkaido Ishikari, newly created by Emperor Meiji. This hatchery was followed by another at Chitose in 1889, which ten years later, added an eyed-egg packing plant to make it the central national hatchery for distribution of salmon eggs throughout the country.

The news that the government was embarking on a program of trout culture was taken by many as a seal of approval for this new technology from overseas. National interest turned quickly to other species, and many new farms were built for the culture of the traditional carps. The first eel farm was built by Kurajirou Hattori, also near Tokyo that same year. The eel was a popular choice, because it was hoped that farming might provide an alternative source of supply. The eel had aphrodisiac connotations, and it was a high-priced delicacy always in short supply. Although eel farming was little more than the organized collecting of young glass eels, or elvers, and fattening them in ponds, the practice spread rapidly around the prefectures through the work of Hattori. Others quickly joined in. Hikotarou Terada started in 1879, followed by Sennemon Harada in 1891, and by the end of the century, artificial eel ponds (some of which were quite large) were supplying over 80% of the total eel production in the country.

(a)

(b)

Figure 7.1 Japan, 1877; the first hatchery built near Tokyo for imported rainbow and brown trout: (a) the Nanaimo building; (b) fish trap.

Figure 7.2 Japan, 1878; the first experimental hatchery for salmon at Kairakuen, Hokkaido (built on the recommendation of an American canning engineer).

Figure 7.3 Japan, 1899; the Chitose salmon hatchery, Hokkaido.

Figure 7.4 Japan, 1909; egg packing at the Chitose hatchery.

As part of its economic and social development program, the Meiji govern-ment built a network of lighthouses around the coast, new industrial ports, and many large and small fishing harbors. By the end of the century, Japan had become one of the first countries to establish a number of regional (prefectural) research stations in support of its national fisheries and had an educational sys-tem that also sent students on overseas study tours. These stations, which were built around the turn of the century, also became the focal points for fish culture, and in particular, experimental work on the propagation of the eel to produce elvers for the farm. The first station was built at Issiki-cho in 1894 to serve the Aichi Prefecture, and this was followed rapidly over the next dozen years by stations in Fukuoka (1898), Mie (1899), Shizuoka (1903), and a smaller district station at Kuwana, again in Mie (1907).

The Meiji strategy of building solid infrastructures for its industries would guarantee the country's leading position in world fisheries and aquaculture for a century. However, this would not be enough without the enabling reforms for farmers, and the government's research and development program for fisheries was complemented by one of the first pieces of land reform legislation in direct support of aquaculture. This was the Reclamation Subsidy Act, which specifically designated large areas of land for reclamation and for the construction of ponds for fish culture.

With the new ownership of the land, backed by legal deeds, the land-values were reassessed with more modest taxes to be paid no longer in kind, but in cash. Such reforms encouraged capital investment in a large number of new business enterprises and in anything other than agriculture. One of the popular enterprises was farming eels, which in turn only increased the demand for more elvers. Unfortunately, the demand was exacerbated by some agrarian uprisings that further reduced the value of land. For a brief period, this encouraged many of the new land-holders to convert their traditional agricultural lands into what were to be more profitable fish culture ponds. As a result, the price of elvers rose to new heights, justifying more experimental research work on attempted breeding (without any success) together with feeding and general husbandry of elvers before distribution to farms for grow-out. Because of the scarcity of elvers and the small size of land-holdings in Japan, the fishponds were small, which fortuitously made them far more manageable and productive.

A technical revolution also took place in the oyster industry in Japan in the 1920s. After some three centuries of raising oysters on nets strung between bamboo poles pushed into the sand, Hidemi Seno and Jyuzo Hori began trials in the Inland Sea to grow oysters hanging from floats. With the greater efficiency in the use of space, oyster production expanded, together with the size and construction of the floats. The oyster hanging-culture technique was also used to raise one-year-old spat with high efficiency, with the result that production tripled in little more than a decade.

By 1912, the year Emperor Meiji passed away, modernized Japan had become a significant world power. It had also continued its historic preoccupation with China over the Korean Peninsula and was then concerned about the imperial interests of Russia and the Europeans in that part of the world. A quick victory in the first war with China in 1895 resulted in recognition of an independent Korea and cession of Formosa and other places strategic to the future of Japan.

Like the European imperial administrators in Asia and Africa, Japanese administrators of the new protectorates pursued activities that reflected the successful policies of industrialization and development in their own homeland. Among many other things, fish and shellfish were liberally shipped from Japan and introduced into the occupied countries. They brought their own Japanese farming experts. For example, in 1910, a fisheries department was structured for the island of Formosa (Taiwan), and a fish culture station was built at Tainan. It was directed by a leading Japanese culturist called Takeo Aoki. He was brought in to work on the propagation and farming of the milkfish, a bony but popular coastal fish first introduced by Dutch occupiers some three hundred years before from Batavia, the administrative center of their widespread colony at the time, the Dutch East Indies. The Tainan station, and others built by the Japanese right through the 1930s, in time became the nucleus of the present Taiwan Fisheries Research Institute. This institute, some fifty years after its founding, would have a major impact on the development of marine shrimp farming throughout Asia.

Japan also had an impact on fish culture in the far east of Russia. The Japanese built the first hatchery for pink and chum salmon at Kalini, in the southwest of the Sakhalin Peninsula in 1925, and operated it until 1939. It proved to be one of the most productive hatcheries ever built and provided a wealth of research data for future management of the Pacific salmon fisheries by the two countries. By the early 1960s, there would be twenty hatcheries on the peninsula and five

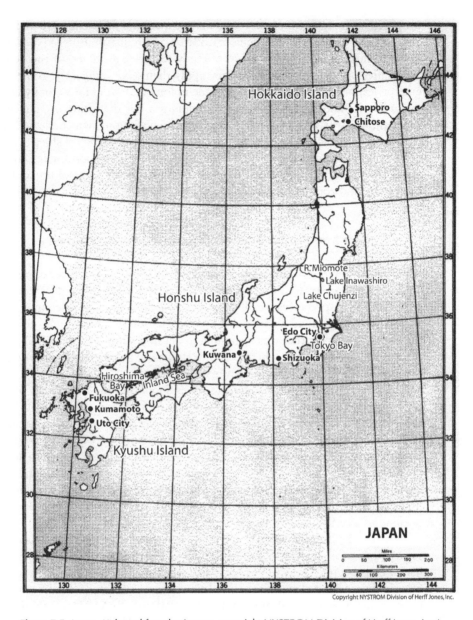

Figure 7.5 Japan. (Adapted from basic map, copyright NYSTROM Division of Herff Jones, Inc.)

more on the disputed Kuril Islands. In 1972, the countries conducted the First Symposium on Aquaculture in the Pacific Ocean in Tokyo.

The strong traditions of fish culture in Japan, together with all the great national changes, continued to have other consequences for the country's aquaculture industry in the long term. The first half of the twentieth century had witnessed the building of individual technical and artisan skills, but the commoners were still held back by a cultural system subservient to the emperor and the ruling classes. However, with the amendments to the Meiji Constitution proposed by General MacArthur and his occupational forces in 1946 and 1947, and with the introduction of democratization and redistribution of lands, thousands of farmers and fishermen were ready to take advantage of the new opportunities offered. Not only did they apply themselves diligently to their own enterprises, but they formed strong trade unions that would eventually control the industries and the lives of their communities. For the future of aquaculture in Japan, this was very important. The fishermen's unions, which virtually controlled the coast, saw aquaculture as a natural extension of their fishing activities and simply integrated it into their system, because it was good for the community. Consequently, almost anything became possible for aquaculture along the narrow coastal zone of Japan.

7.4   Conservation and compensation measures, North America

Although the interest in hatchery propagation of marine fish in Europe and North America was rapidly declining by the 1920s due to the lack of results, work with freshwater fish species, especially the Pacific salmonids, was becoming highly successful in the United States and Canada. The success was timely, because it pre-empted the crisis that was beginning to develop in all the salmon fisheries of the Pacific Northwest. Nowhere was the crisis more keenly felt than throughout the broad area of the Columbia River watershed.

The fisheries of the giant Columbia River supported the livelihoods of some fifty thousand persons and yielded about eighteen million pounds of fish each year. However, with the opening of the Pacific Northwest territories at the end of the century, the hydroelectric dams, the diversions of water for irrigation, poor logging and mining practices, and many other factors all helped to degrade the valuable salmon resources of the basin. Perhaps the greatest demonstration of the dramatic impact of such activities was the wiping out of the Celilo Falls fishery, estimated to be about 2.5 million pounds of fish, by the completion and opening of The Dalles dam in 1956.

Even before the first dams were built, and when Oregon and Washington were still territories on the verge of statehood, overfishing of the Pacific salmon had already begun. The invention of the steel canning process opened the coastal fisheries to accelerated exploitation. Salmon canneries located in the Washington territory in 1866 packed 4 thousand cases of salmon. By 1883, the number had reached 629 thousand cases, and then dropped to 321 thousand cases in 1889, the beginning of the great decline. This decline did not go unnoticed. In 1895, the young state of Washington appropriated $20,000 for the first hatchery on the Kalama River, a tributary to the lower Columbia. By the turn of the century, many millions of fish were being released from hatcheries.

The states of Oregon, Washington, and Idaho, and numerous federal agencies, currently share responsibilities for the anadromous fish runs in the Columbia and Snake river systems, which drain over 670 thousand square kilometers of the region. At the start of the 1900s, about fifteen million adult salmonids migrated each year from the sea, ascending the Columbia far into Canada and up the Snake River through to Wyoming. Now, at the beginning of the third millennium, fewer than 2.5 million adults make this journey. They are stopped below the Grand Coulee and Chief Joseph dams south of the Canadian border, and by Hell's Canyon Dam on the Snake River, where fish passage facilities end. There are nine dams on the main stem of the Columbia, and four on the Snake with fish passage facilities. At least $60 million to $70 million are spent each year by federal agencies to try to halt the decline of anadromous salmonid runs, supplemented with funds derived from the sale of hydroelectric power and irrigation water.

Many fish hatcheries and fish ladders were constructed in the region to mitigate some of the losses, but it was not until the Federal Power Act of 1920 that fishways were required at all private power projects. The Fish and Wildlife Conservation Act of 1934 that followed was the most important legal authority for ensuring both protection and compensation for the salmon fisheries affected by federal water projects. Together, these two acts would be responsible for over $400 million spent on fish passages and hatcheries constructed in Columbia Basin power and water projects at the time.

Serious hatchery expansion did not begin until the 1930s, after construction of the Grand Coulee Dam on the upper Columbia and Bonneville Dam on the lower Columbia, the first low-head dam with fish passage facilities, and the first that adult fish encounter as they return from the ocean. But the early hatcheries built to compensate for the Grand Coulee Dam did not meet expectations: that is, the benefits to the fisheries were not readily apparent, as had been the case with marine hatcheries, and therefore, specific compensation for individual projects could not be readily identified. Consequently, the Mitchell Act in 1938 authorized appropriation of federal tax revenues to restore and enhance the salmon resources of the Columbia Basin as a whole.

Although all the funds appropriated by federal legislation for the compensation program were not completely released until the end of the Second World War, they enabled a massive scaling-up of hatchery and propagation facilities. This enabled a substantial technical and scientific research-and-information base to be established in the region on almost every aspect of the biology and culture of the five key species of Pacific salmon. The knowledge and experience would be significant factors in the farming of both North American and European salmonids thirty years later.

One such major technical breakthrough in the late 1950s was the development of pelleted feeds. These artificial feeds were correctly formulated with ingredients supplemented with vitamin and mineral additives. They were very palatable, especially the Oregon moist pellet. They were also pasteurized, which allowed long storage, at least for months, and which prevented the spread of disease through contamination. They also increased survival and fitness of juveniles, and through better feed conversion, reduced the costs of producing smolts. In addition, they could be formulated with new antibiotics for disease control when it was necessary. Subsequently, the feed formulation and processing technology developed for the salmonids in North America in the 1950s became the basis of all feeding technology for all aquaculture in the years to come.

As each successive dam was constructed, mitigation hatcheries followed. Between 1960 and 1976, two billion juvenile salmon and steelhead trout, weighing over twenty-seven thousand tons, were released from eighty-one hatcheries and supplementary rearing ponds. By 1976, the annual release of juvenile salmonids in the Columbia Basin was 54% of all cultured juvenile anadromous fish released in the entire Pacific Northwest. These hatcheries became extremely successful, and some had total recoveries as high as 12% for fall chinook salmon. Hatchery stocks of Columbia Basin chinook and coho salmon ranged as far south as northern California and as far north as southeast Alaska, which encouraged many small coastal towns to build fleets of charter vessels for anglers. The fish were also intercepted offshore by growing numbers of commercial trollers and in the rivers by gill-netters.

There is no doubt that North American fish culture technology developed in the first half of the century was effective, and that it was the foundation of the future growth after the Second World War. In terms of meeting the demands for enhancement and conservation, the public hatcheries constructed for mitigation of the losses of Pacific salmon in the United States and subsequently in Canada played a valuable role in maintaining the salmon fisheries both inshore and offshore. However, Pacific salmon was not the only interest in the United States at the time.

At the opposite corner of the nation, far to the south, economic interests lay more in revitalizing depressed agriculture. From the beginning of the century, there had been a growing interest in the construction of farm ponds, which then could be seeded with local warm-water fish to supplement the states' food supply. However, there was little concern for any pond management, because scientific knowledge about such things as water quality, stock density, fish nutrition, and fish diseases was nonexistent. Not surprisingly, as the country moved through the social changes brought about by the First World War and unwittingly toward the Great Depression, the enthusiasm for fishponds on the farms of the country had long since faded. After 1929, this picture reversed again. With ruin and poverty staring many farmers in the face, all marginal agricultural lands were seen as potential areas for the construction of ponds for the farmed production of local warm-water fish.

This time, however, the farmers could not make the same mistakes. Somehow, someone had to determine what lay behind the control and management of these fishponds to make them productive and valuable. An inspiration to meet this need through practical studies in the natural inland watersheds in the plains of Alabama came from Homer Swingle, who had joined the Alabama Polytechnic Institute in 1927. In 1934, with a small team of scientific colleagues, Swingle built ponds at the Alabama Agricultural Experiment Station and studied the principles of fishpond management for reliable annual production. The series of publications on the observations and methods became landmarks that would serve the postwar generations of fish farmers in the country. By 1943, the Experiment Station had constructed over five hundred hectares of ponds throughout the entire watershed near Auburn. By 1952, there was the almost unbelievable number of 1.7 million farm and ranch ponds throughout the country.

The Polytechnic Institute, in time, became Auburn University. Its rudimentary fish farming activities that it had inherited were the foundation of one of the most important freshwater fisheries programs in the United States, which would have lasting consequences. It was the basis of most of the technical and practical knowledge regarding farm pond management in the country, which would help freshwater fish farming become a major industry. It also initiated some of the first work on channel catfish production, which would later surpass 180 thousand tonnes annually and become a valuable regional economy. It was a center of scientific research, which would eventually expand far beyond the study of locally important species of fish. It was a focal point for the education and training of future state and federal administrators, as well as of research scientists. Subsequently, it would repeat almost all these activities at the international level and became one of the world's leading institutions in fish farming.

7.5   From small seeds grow

With some specific exceptions, the failure of fish farming to capitalize on the excitement generated by the International Fishery Exhibition, and to make the complete transition from research and development into a viable economic industry between 1900 and 1950, must rank as its greatest lost opportunity. The explanation was simple. Apart from the needs of mitigation and conservation, there was no real need for farm-raised products in the commercial markets. World population was still relatively small, with less than two billion people at the turn of the century, and still well below three billion in 1950. Fish and shellfish were plentiful and cheap in the developed world, and there was no centralized international voice responsive to the growing needs of the developing world.

The first half of the twentieth century was therefore a period of sporadic scientific and technical events, many of which would take on new significance in the second half. One small event stands out above the rest, because it was subsequently a major catalyst for much of what followed.

In 1939, a marine scientist called Gunnar Rollefson, working in a government research institute in Norway, discovered that the nauplius of the brine shrimp was a useful live food and small enough for feeding marine fish larvae, such as the plaice and plaice-flounder hybrids he was rearing in his laboratory. It was also a live food that could be prepared easily when it was required. These crustaceans had masqueraded for years in folklore under the name of "fairy shrimp," because they always seemed to appear out of nowhere, as if by magic, in saltmarsh pools or in manmade salterns that previously had been dry for months on end. But the tiny nauplii had actually emerged from eggs that had been carefully encysted by the female adults when they were facing drought and death.

That is, fertilized eggs can be retained in the female shrimp's ovisac until a tough chitinous shell or cyst has formed around each of them. The encysted eggs can either hatch within the ovisac, where the embryos develop into nauplii before they are released, or they are deposited in the water, where they will hatch if conditions of salinity and temperature are favorable. In unfavorable conditions, the protected eggs, which are of the color and size of fine grains of sand, remain dormant for long periods without destruction of the germ cell as they lie in the salty, dry substrate, or are blown in the wind until brought back to life by immersion once more in warm salt water.

When Rollefsen collected dry, encysted eggs and incubated them in warm sea water for twenty-four hours, each one yielded a small pink nauplius, which he could collect by the organisms' attraction to light. With this simple discovery, Rollefson not only found a suitable live food that enabled the subsequent culture of a host of marine species, but one that could be stored conveniently in a tin for years, and only taken out and hatched when required.

Chapter 8

Aquaculture in a World at War (1935–1945)

Abstract

Fish culture, not silenced during the Second World War, achieved milestones in Scotland, Africa, the Middle East, North America, and Eastern Europe—some long-lasting, others dead-ended. Fertilizing Scottish sea lochs increased flatfish biomass; natural predators ate the fish, but the work was abandoned before economic analysis. South African mines, expanded for European war production, fed workers pond-cultured tilapia so successfully that the fish farms survived political upheavals through the 1980s. European emigrants' traditional common carp culture evolved into skilled Israeli kibbutz polyculture systems, making Israel self-sufficient in fish production. Nuclear development's environmental research on the Columbia River led to Lauren Donaldson's ocean ranching of salmonids at the University of Washington. Selective breeding experiments produced superior fish and larger returns. Postwar Communist fish farming cooperatives' culture of species introduced centuries earlier was successfully expanded in war-devastated Central Europe. The Soviet Union's collapse demonstrated cooperative economic problems, and triggered failure of Soviet-controlled Caspian Sea sturgeon ranching due to poaching.

8.1   Introduction

The first half of the twentieth century closed with the world at war. Fish culture, however, was not to be silenced during this period. In fact, during the years of the Second World War, there were a number of singular events that were milestones in the painfully slow development of aquaculture through the four previous

The History of Aquaculture. By C. E. Nash. Published 2011 by Blackwell Publishing Ltd.

decades. Some of these events had long-lasting significance; others would become dead ends. One took place in the remote sea lochs of western Scotland, another in the rich copper mining belts of Central Africa, a third in the arid lands of the Middle East, and the fourth in a top-secret site in a remote northwest corner of the United States. The fifth and most significant event was the direct result of the war's aftermath. This was the complete reorganization of fish farming in the Eastern Bloc countries under their new Communist regimes.

8.2   Fertilization of a Scottish sea loch

The war years from 1939 to 1945 prevented everything but nearshore fishing around the countries of Europe. The fishing grounds of the North Sea were closed, and the fishermen of most European countries had themselves joined their respective naval reserves. On the markets, marine fish were in short supply, but the demand for fish always remained. For European nations dependent on fish, and with reputations as fishermen, this meant that a number of alternatives were inevitably explored.

In the British Isles, nearshore fishing was suddenly concentrated along the safer western coasts, but at a much reduced level. A group of marine scientists suggested that nearshore fish production could be improved by enhancing the productivity of some of the sea lochs of Scotland with inorganic fertilizers, and by increasing the resources of fish through transplantation of juveniles, particularly flatfish, to these prepared sites. The idea was not without merit, and it was not in fact their original concept. Similar, but albeit much smaller fertilization experiments had been tried in a small basin on Saelo near Hardanger in Norway in 1908 and repeated again in Norway in 1931 and 1932 to improve production of oysters.

The first experiment in Scotland began in 1942 in Loch Craiglin in Argyll under the direction of Fabius Gross from the University of Edinburgh. He gathered about him a number of top young marine scientists of the day, many of whom in later years were destined to become the directors of the nation's leading marine science laboratories. In spite of the pressing need for fertilizers by a country desperately trying to produce all its own food, Imperial Chemical Industries Ltd., the nation's largest national producer of fertilizers, was permitted to supply Gross and his team with sodium nitrate and commercial super-phosphate. The company also supplied Gross with some modest research funds, but for the most part, the achievements of the program were due to the cooperation of marine scientists in all the principal institutions of the country, sharing or lending their precious equipment in times that were very austere.

The precious fertilizers were distributed evenly, but with some difficulty, over the eighteen acres of Loch Craiglin at regular intervals from 1942 through to the end of 1944. From the data collected by the scientists, it was clear that they greatly increased the productivity in the loch. The nutrients were rapidly converted into a feeding ground for fish that approached, although on a small scale, the richest natural feeding grounds known. Levels of zooplankton and larger food organisms suitable for growing flatfish were also increased, but not with the consistency required to support large populations of fish. At times, the growth rates of the young fish were good, particularly those that had not been stressed by marking with small discs, and much better than those of the control population in nearby Loch Sween. By 1943, there was also noticeable growth in the numbers of predators working the shallow waters of the loch, particularly eels, cormorants, and herons, and they systematically removed practically all the larger fish by the end of the year.

However, the economics of one of the first true attempts to farm the sea was never put to the test. More research was obviously required, but this became less important. The Battle of the Atlantic was over by 1944, and the theater of war in Europe was grinding to its end. Fishing vessels were beginning to return to the seas in earnest. Both fish and shellfish once again became more available on the local town markets, and tinned salmon was a luxury to be found in food parcels from the United States or on a shelf in the local grocers' shop. The scientific group working on the Loch Craiglin project was disbanded, and perhaps all further technical interest was lost with the untimely death of Fabius Gross in 1947.

8.3   Pond construction in Africa

Another singular event in the years of the Second World War heralded the era of massive pond construction in Africa. Africa was a major resource of a spectacular array of raw materials for the Western Allies, and the rich mines of South Africa, Northern Rhodesia, and the Belgian Congo were expanded and exploited to supply the war machines of Europe.

Large pools of labor were brought in to work the mines of Africa, to build the railways, and to man the ports to keep up the delivery of copper and iron ores. This large labor force had to be housed and fed around the clock. For a period of time, their food, and particularly meat, was supplied by the large hunting parties that were employed by the mine owners for this particular purpose and by deliveries of beef from South Africa. However, it was not long before the herds of wild game close to the mines had been depleted, and the tinned beef from South Africa was directed toward the armies and hungrier markets of war-torn Europe. Meat for the hard-working miners was becoming a problem again. Because both Great Britain and Belgium had been actively financing development of fish farming of tilapia in their respective colonies in Africa for several years, their administrators came up with the idea of a joint war effort and devised a program for the construction of ponds to produce fish to feed the transient miners and their families.

Farming the native tilapia in Africa had been tried for the first time in Kenya in 1924 by the British Colonial Service, and it had spread quickly to other neighboring East African countries. In 1937, it was introduced by the Belgians to the Congo. Pond production of tilapia was clearly shown to be successful almost everywhere it was attempted, provided that the ponds were regularly managed. Consequently, in Northern Rhodesia in 1942, many thousands of small fishponds were dug and stocked with tilapia and other native species for food for the miners. With the necessary vigilance and discipline dictated by the circumstances, the project succeeded.

The project succeeded so well, in fact, that at the end of the war, the two colonial countries decided to continue. In 1949, they organized the "Conference Piscicole Anglo-Belge" in Elizabethville for further postwar development of fish culture for food production. This then led to regular initiatives for further pond construction programs with tilapia. They started in Rhodesia (1950), and followed with further conferences in Entebbe (1952), Brazzaville (1956), and Lusaka (1960).

Oddly enough, these same small fishponds built in 1942 in the mining belt of Northern Rhodesia would surface again some forty years later. For a brief spell in the 1980s, when discipline returned to encourage private investment by small-scale farmers in the country by then renamed Zambia, farm production of tilapia in these ponds reached new heights. Unfortunately, it would then be thwarted all over again, as political and economic problems once more engulfed the country.

8.4   Farming the arid lands of the Middle East

The first fish farm in Israel was developed by immigrants from Central Europe in 1934. Into what was then still Palestine, these new settlers brought with them the experience for the culture of their traditional and prized fish, the common carp. The idea quickly spread among the settlers, and many small private fish farms were started wherever there was available water. Although these independent farms were widely scattered around the country, the majority found it difficult to survive. There was intense development in the region at the time, and a strong communal demand to share the limited resources of land and water. However, communal sharing was a fundamental characteristic of the *kibbutzim*, and was proving to be an appropriate and successful approach for food production for the immigrants and settlers.

The first kibbutz fish farm was established in 1938 at Nir David, which decades later would become the major national research center for fish farming in the country. The numbers of farms around the country grew in response to the increasing demand for freshwater fish through the austere years of the Second World War. By the end of the war, there were thirty productive kibbutz farms, occupying an area of six hundred hectares and producing well over a thousand tonnes of carp. About half the farms were in the Beisam Valley, where the water was slightly saline; the rest were in settlements in the Jordan Valley, north of the Dead Sea, and spread down the coastal plain south as far as Tel Aviv.

At first, the early culturists used the traditional methods that they had brought with them from their homes in Europe, but as land and water became scarcer, even for the kibbutz, they developed new systems and practices of their own, which were unique and highly productive. They imported Chinese carps and tilapias, and with a detailed understanding of some principles of fish biology and environmental behavior of these species, they developed highly skilled systems of polyculture that eventually would make Israel self-sufficient in fish production, at least for one period in its future. After one or two years of production, the country even was able to export fish.

8.5   Fish and fission

In 1942, in an old squash court at the University of Chicago, Enrico Fermi succeeded in sustaining a fission chain reaction. This was a process critical to the success of the Manhattan Engineering Project, the Allies' solution for a rapid end to the continuing years of war. Fermi's reactor was the prototype for five production reactors manufacturing plutonium, the isotope compressed into the first atomic bomb tested in New Mexico. These reactors were built at Hanford, a remote location in Washington State, on a site at which enormous volumes of cooling waters could be tapped from the great Columbia River. Eager to learn everything about the ramifications of the radioactive isotope before it was liberated from the bomb, the director of this top-secret site (known as W) needed biologists, and specifically aquatic biologists, to undertake some very specific studies on the effects of high water temperature and contamination on the native flora and fauna. He was directed toward an assistant professor at the University of Washington, Seattle, called Lauren Donaldson.

Already a mature research student, Donaldson joined the School of Fisheries at the University of Washington in 1932 to pursue his studies on fish nutrition and to concentrate on selective fish breeding and growth. When the United States was finally drawn into the war at the end of 1941, Donaldson was teaching fisheries biology at the School of Fisheries and passing summer vacations on long field trips in the upper Fraser River system in British Columbia. His missions were on behalf of the International Pacific Salmon Fisheries Commission in their quest to rehabilitate the area beyond the great Hell's Gate Dam. As he was about to set out for the field in 1943, he was diverted by an urgent cable sent to the Fisheries Commission at its Canadian headquarters, ordering him immediately to the Office of Scientific Research and Development in Washington, D.C. There, he was interviewed by the chief of the Medical Section of the Manhattan Engineering District, a name that of course meant nothing to him. Eventually, he was asked whether he would undertake some very special studies on fish and other aquatic organisms, to which he readily agreed. Returning to Seattle, he found himself director of what was called the Applied Fisheries Laboratory. Some of his equipment had already been delivered and was ready to be unpacked.

Early in 1945, a second laboratory was provided at Hanford, and even though the war came to its abrupt end in August of that year, Donaldson and his team in the Applied Fisheries Laboratory spent four more years on research associated with the nuclear test experiments then being carried out by the newly formed Atomic Energy Commission. It was on the long, 250-mile trip from Seattle through the Cascade Mountains to Hanford, and up and down the Columbia

River, that Donaldson developed his ideas for conditioning salmon to return from their migration to where *he* wanted them to be. He believed that the approach should work, because he had already had some evidence from his early experiments for the Fisheries Commission in British Columbia. The opportunity for him to test his theories came in 1948, when plans were drawn up for the new University of Washington Fisheries Center on the shores of Portage Bay on Lake Union, which was linked directly to the sea by canal. Donaldson's experimental rearing ponds and a salmon ladder were finished long before the classrooms were opened, and in 1949, twenty-three thousand juvenile chinook salmon were released to find their way out to sea. Twenty-three would return in November 1953.

Within the decade, Donaldson's hatchery had runs of chinook, coho, and sockeye salmon through the front door of the university, and the runs would become well established. His salmon were famous as the only research animals that returned to the laboratory and the classroom of their own volition. This unique attraction, along with his subsequent research, would help put the School of Fisheries in the forefront first of fisheries and then of aquaculture education in the United States for the next thirty years.

Firmly established at the University of Washington, Lauren Donaldson had the facilities and resources to do other research that had long-lasting benefits for the renewed interest in aquaculture that was just around the corner. He began to work with rainbow trout, and through years of selective breeding and hybridization, he produced a fish far superior in size than any wild one. The "Donaldson super trout" became famous and proved to be very popular among the growing numbers of trout farmers, not only in the United States but all over the world.

8.6   State-operated fish farming in Eastern Europe

Long before the end of the nineteenth century, fish farming in Eastern Europe was quite well established. The medieval feudal system of fishing, in which a right to fish was given or withdrawn by royal prerogative or whim, had slowly been eliminated, and by the end of the nineteenth century, many of the countries had enacted legislative reforms regarding the use of land and water, the right to fish, and the right to farm fish in fresh water. Consistent down through the ages, however, was the fishing cooperative. Initially, its purpose was to bind together the fishermen who were needed to set and haul the large seine nets that were operated from the shoreline of inland lakes and ponds. Because the system was effective for management, it frequently remained long after the rights for individuals were established.

Fish production in the large land-locked areas of Eastern Europe was predominantly carried on the backs of the warm-water species that had been introduced centuries before. By the end of the nineteenth century, some of the farmers were beginning to eye the brown trout and rainbow trout systems that were attracting their colleagues in Western Europe and North America.

Despite the long tradition in fishing and fisheries enhancement by the East-erners in the lakes and river systems of the broad interior of Europe, for the first half of the twentieth century, fish farming in general did not fulfill its projected potential, in spite of repeated government initiatives over two or three decades. In the Eastern European countries of Czechoslovakia, Hungary, and Bulgaria, for example, the cultivation of warm-water fish had been included in the basic curricula of their many agricultural schools. In Bulgaria, the first demonstration ponds for farmers were built at schools in Sadovo and Russe as early as 1892, and they are still being used to fulfill their original purpose a century later. The first rainbow trout farm was located in Gabrovo, and a large state-owned trout hatchery was built at Samakov to provide Bulgarian farmers with free seed until they were established with their own broodstocks and seed resources. However, the scheme was never well promoted and few farmers took advantage of it. Twenty-five years later, fishponds covered no more than forty-two hectares of private and state-operated farms, and production remained abysmally low.

In Czechoslovakia, land drainage policies for agriculture and animal hus-bandry meant the abolishment of many of the traditional ponds, drastically affecting the national economy. A minor resurgence did take place at the end of the First World War, with the Declaration of Independence in 1918. A general nationalization of resources included the majority of the fishponds. Together with reintroduction of traditional dress for the different grades of fishers and a return of the old ceremonies and festivals, there was new infrastructure. At Vodňany, a fisheries school was built in 1920, and then the Research Institute of Fisheries and Hydrobiology in 1921. Another field research station was built out at the old Lnare ponds, famous for their strain of blue carp. At the same time, there were also some favorable land reforms, which brought more ponds under state control. However, all these progressive moves were nullified when many of the old ponds proved too costly to refurbish, and the government introduced price controls, which in most cases, hardly covered production costs.

In Hungary, things were much the same, although the outcome was a little different. A research station was established in Budapest in 1906. It was called the Royal Experimental Station for Fish Physiology and Wastewater Purification. In time, the institution would become the Fish-Culture Research Institute, famous for its pioneering work with fish-*cum*-duck farming at Szarvas. The station got into duck farming somewhat through a back door. When it was first established in 1969, the station was financed under a ministerial program for increased meat production, which required work with ducks and pigs, as well. Naturally, the station picked up on the integration of all three into very productive and profitable systems.

However, none of the national efforts to build infrastructure in the 1920s for the projected growth of the fish-farming industry in Europe was enough to encourage many agricultural farmers to try their hand at aquaculture. One good reason was that the markets were predominantly national. There was fierce competition for the small but traditional export markets for warm-water fish in Germany and Austria, a competition that was eventually won by the independent fishermen working the natural lakes and rivers in Hungary and northern Yugoslavia. They could catch fish much more cheaply than the farmers could produce it. But the fishermen's victory did not last long, when the 1930s were suddenly affected by the economic crisis that echoed all over the world. Consequently, with little real enthusiasm for farming, the owners left many of the fishponds to overgrowth, after which the ponds dried up.

The fish farmers in the Eastern European countries who did make it through the 1930s lost everything when most of their properties and farms were destroyed in the battles of the Second World War, which ranged around them for six long years. And when the war finally ended in 1945, the ground rules had all changed. Under the new Communist regimes, traditional fishing rights were given to newly established fishermen's cooperatives, and state-owned fish farms were constructed to operate within the newly formed network of agricultural state farms. Any remaining private farms disappeared.

This compulsory reorganization of the agriculture and fisheries sectors in the postwar Eastern Bloc countries proved to be an effective mechanism to develop and expand the number of fish farms very quickly. In Bulgaria, for example, when the regime finally collapsed in 1989, there were more than three thousand hectares of ponds spread throughout the seven hundred state-owned agricultural cooperatives or standing as individual state-owned fish farms. By 1984, there were some twenty-three thousand hectares of pond farms in Hungary. In Czechoslovakia, there were over fifty thousand hectares of ponds, many of which were still the large ponds constructed centuries before.

With the collapse of the regimes in all the Eastern Bloc countries, the transfer from state-owned cooperatives to private enterprises did not go smoothly. Many farms quickly became bankrupt. Although the workers had the technical skills, they had not the money to buy the basic farm resources that had been subsidized by the government for decades. Furthermore, they were inexperienced in selling the fish profitably once the regulated and noncompetitive market of the former Soviet Union had been lost.

The same long-term scenario also was taking place inside the Soviet Union itself. One of the postwar priorities for the Ministry of Fisheries was the construction of state fish hatcheries for sturgeon to enhance the natural fisheries, especially those of the Caspian Sea and in Siberia. With the construction of about ten hatcheries around the deltas of the Volga River and the Ural, the tightly-regulated Soviet fishermen were soon taking about 90% of the catch from the Caspian, and the Ministry of Fisheries was in firm control of the global market for caviar. The rest of the Caspian fisheries had long since been traded away by the Russians to Persia (Iran) in 1828. But with the collapse of the Soviet Union in 1989, and the new legal framework for the five states that then surrounded the Caspian Sea, it was immediately open season on Caspian sturgeon. Only the Iranians have maintained a tight control on the restocking and harvesting. The peak of the sturgeon harvest from the Caspian Sea was thirty-nine thousand tonnes in 1915. In 1988, the year before the Soviet Union was disbanded, the annual catch was steady at about twenty thousand tonnes. Five years later, it stood at a bleak seven thousand tonnes, with few prospects for recovery in the approaching new millennium.

Chapter 9

Postwar Pioneering (1950–1970)

Abstract

Post-Second World War, interest in fish farming declined, and wasteful fishing returned. Monitoring passed to the United Nations Food and Agriculture Organization. Agriculture included fisheries and marine products. Technical development paralleled social development. Prewar Japanese research led joint postwar aquaculture development. Oyster hanging-culture in Asia and French oyster industry recovery from wild and hatchery seed developed alongside Spanish and Dutch mussel and oyster farming. Britain's White Fish Authority prewar phytoplankton research developed oyster hatcheries and flatfish farming, and ultimately, Russia, France, Italy, and Greece's national aquaculture programs. Artemia research was key to fish hatchery development. Catfish and crayfish farming became North America's industry; hatchery developments at Milford, Connecticut, restored oyster stocks, creating today's cultchless oysters. America and Norway pioneered saltwater net-pen salmonid farming. Stratton commission's Blue Revolution founded the National Oceanic and Atmospheric Administration, whose Sea Grant Programs led freshwater prawn and gray mullet farm development.

9.1 The creation of the United Nations Food and Agriculture Organization

Despite the increased interest in the expediency of fish farming for food production in the years of the Second World War, the end in 1945 saw an immediate return to the old wasteful ways. Buoyed by almost ten years without extensive

The History of Aquaculture. By C. E. Nash. Published 2011 by Blackwell Publishing Ltd.

harvesting, coastal and oceanic fisheries resources were back to strength. The world harvests of fish and shellfish increased steadily as old fleets of traditional fisheries nations were modernized and expanded, and new fleets of newly in-dependent countries entered the fisheries at will. By 1950, the world harvest of 1938 (about 20.5 million tonnes) had been regained and soon doubled.

Monitoring of the world fisheries' catch passed from the old International Institute for Agriculture, which had been based in Rome since its inception in 1905, to the newly formed United Nations Food and Agriculture Organization (FAO), located in Washington, D.C. Under the League of Nations, the Interna-tional Institute had been a clearinghouse for world agriculture information and a forum for major issues. Its technical arm had been more concerned with prime agricultural commodities (cereals, dairy products, and meats), forestry products, and a group of oddities, such as olives, wine, and honey. The Institute had not been at all interested in either fisheries or fish culture other than maintaining statistics on world fisheries and fish trade. Library records showed only two reports on fish culture: a survey of carp breeding in rice fields in Italy in 1913 and of fish and fish culture in Hungary in 1916.

The new FAO, on the other hand, firmly stated in its founding constitution that it interpreted "agriculture" to include "fisheries, marine products, forestry, and forestry products." Thus, when it moved from Washington to Rome at the beginning of 1951, the organization created a fisheries department and an annual program of work. Fish culture was included in its overall purview, and staff was appointed. S.Y. Lin, the superintendent of fisheries research for the Hong Kong government, was hired to be the first fisheries biologist with special responsibilities for freshwater fisheries and "pond fish culture," as it was called at the time, to be quickly joined by Shai-wen Ling from Taiwan and Walter Schuster, a Dutch biologist who had been working in Java (Indonesia). It was Schuster who, on behalf of the Dutch administration, had organized one of the first meetings for inland fisheries experts in Surabaya in 1939. The subject was the potential of the tilapia, which had mysteriously appeared in East Java earlier that year. FAO also established a number of regional offices around the world and began the formation of its regional fisheries commissions and councils. As a result, an international development structure was put in place that could assist its member nations to exploit both inland and offshore resources, and to take advantage of the state-of-the-art fish culture.

Because of the strong tradition of fish culture in Asia, the new FAO regional office for Asia and the Pacific, which was moved from its original home in Singapore to Bangkok, rapidly became active in the field, together with the newly formed Indo-Pacific Fisheries Council. The council produced many technical reports on brackish-water fish culture in the proceedings of its early annual meetings held throughout the region, and the regional office organized its first international seminar and training course in Djakarta and Bogor in 1951 for eighteen students from the region.

The experts hired by FAO, aided by other fisheries officers still employed in colonial administrations, were greatly in demand to conduct national surveys of resources for fish culture and to create projects for their development. One of FAO's first projects was to assist Haiti to meet its acute food shortages, particularly that of animal protein, through the culture of tilapias and carps. The project was under the direction of Shimon Tal from Israel. It began in 1950, and international assistance would last in one form or another for the next forty years, with little evidence of any real contribution to the ongoing problem. A similar project began in the Dominican Republic in 1953.

FAO experts were also behind the project to spread tilapia species throughout the Asia and Pacific region in an attempt to introduce fish as readily available protein for the rural poor. It was intended that communities could obtain subsistence by harvesting the easily grown fish from their local watersheds or in village ponds. Lamentably, the introduction of tilapia was to have long-lasting repercussions, because the ubiquitous species rapidly replaced much of the indigenous fauna and became a pest wherever it was released.

However, FAO soon demonstrated its forte for gathering information through the organization of international meetings. In May 1966, the expanding Fisheries Department called on experts to gather in Rome to attend the "World Symposium on Warm-Water Pond Fish Culture." This was the first meeting of its type and was attended by delegates and technical representatives who would soon become the household names in the growing field of aquaculture. They presented information on the status of fish culture in their regions or individual countries, and some gave technical papers. The proceedings are a compendium of postwar activities in fish culture. The symposium was soon followed by another in Mexico City in 1967. This time, the gathering was heralded as a world scientific conference on the topic "Biology and Culture of Shrimps and Prawns." The proceedings show that the meeting was largely on the biology of the species, because with the exception of the work in Japan, the farming of crustaceans at that time was still predominantly a figment of the imagination.

Further information on the widespread and productive fish farming in Asian countries was revealed at an international seminar on the "Problems and Possibilities of Fisheries Development in Southeast Asia" in 1968. This meeting, held in Berlin, was sponsored by the German Foundation for Developing Countries and FAO, and brought together a large mix of fisheries and aquaculture experts for the first time. Instrumental in organizing the conference was Professor Klaus Tiews, Director of the National Institute for Fisheries in Hamburg, who would remain a primary proponent of aquaculture in Europe for another twenty years. In particular, Tiews was responsible for much of the positive attention on the emerging aquaculture sector in Europe by one of FAO's most successful offspring, the European Inland Fisheries Advisory Commission.

9.2   Aquaculture in international development

The early postwar years were characterized by an emphasis on basic research in freshwater fish and the traditional shellfish culture. Although many of their territories were rapidly disappearing, the European colonialists, through their overseas civil services or special departments for international cooperation, continued to assist their old colonies. Great Britain, for example, recognizing the worldwide postwar shortage of food, actively promoted fish farming as an adjunct to peasant agriculture "as a means of supplementing the protein diet of the peoples of the Commonwealth." This contributed significantly to the further introduction of science out of the laboratories in Europe and into the fields of Africa and Asia.

Successful scientific application in foreign fields, however, was far from being easy, and great credit is due to the pioneering work of a handful of individuals who devoted the greater part of their lives to working overseas. One such man was Fred Hickling. He was one of the first of many British fisheries scientists who began to concentrate in the colonies of East and Central Africa. There, he studied the genetics of tilapias and produced the first all-male populations. But after travels that took him all over the world, he soon realized that his research ideas could be tested and applied more easily in the countries of Asia that had a long tradition in fish culture. After an extensive search of prospective sites for a regional fish culture research station in Malaysia, North Borneo (Sarawak), Borneo (Kalamantan), Hong Kong, and East Africa again, he chose a place near Malacca on Penang Island.

Hickling proposed a station with about forty hectares of research ponds spread over the hundred-hectare site, and as a trained classical scientist, he included in his plan the only Latin square of thirty-six, one-acre ponds for research ever built. Construction of the Fish Culture Research and Training Institute, as it was called, cost £252,000—a considerable sum in 1951. The bill was paid by the British Treasury through the Colonial Development and Welfare Research Fund. In addition to advancing fish culture and regional training at the Institute, Hickling's research contributed greatly to the early advances in knowledge of soil and water chemistry of fishponds. In 1962, he published his great experiences in the first book ever devoted to fish farming. It was called *Fish Culture,* and because it became the manual for every project manager all over the world, it was updated and reprinted again in 1971. Hickling's fundamental research at Penang was continued by his successors. One of these was James Shelbourne from Lowestoft, who was seconded briefly in 1962 from his pioneering work on flatfish culture; he was followed by Geoffrey Prowse, another of Hickling's colleagues from East Africa.

Antoon De Bont and Marcel Huet spent much of their working lives between the University of Louvain and the Research Centre of the Department of Waters and Forests in Belgium; in Central Africa at the fish mission in Katanga in the Belgian Congo; or at the International Training Centre for Inland Fisheries in Java. They worked together on the breeding of tilapias and carps and carried out research on general limnology relevant to inland fisheries. Huet started to publish his many experiences in his *Traité de Pisciculture* as early as 1952, and his later editions in time became the classic *Textbook of Fish Culture,* first published in 1970 and reprinted many times. Jacques Bard, from the French Technical Centre for Tropical Forests, began his work in West Africa, but later pioneered developments in South America on introduced tilapias and many of the indigenous species. Other well-known fisheries experts of the time included

Wilhelm Schaeperclaus from Germany and Elek Woynarovich from Hungary. Schaeperclaus was probably the most experienced fish culturist of all, having written his *Textbook of Pond Culture: Rearing and Keeping of Carp, Trout and Allied Fishes* as early as 1933. However, it was not translated from his native German until 1948 and did not receive the recognition it deserved.

The opportunities for these international pioneers in the postwar years were many. Interest in fish culture was widespread, and funds were available as part of postwar restructuring policies directed particularly for producing food protein at the rural level. In India, for example, many new commercial farms were established in Assam, and in Madras State, twenty-eight rural fisheries demonstration units were linked in a program of surveying and stocking inland waters and tanks (irrigation reservoirs) with over twenty million fingerlings. Training of national technicians needed for these state projects was begun at the Central Inland Fisheries Research Station at Barrackpore. Similarly, in Ceylon, efforts were made to revitalize the windowpane oyster industry and to introduce new species for the enhancement of inland waters and tanks. In Thailand, the demand for fingerlings by new fishpond owners around Bangkok could not be met by the resources of the national Pond Culture Stations at Bangkhen, Borapet, and Kwan-payoh, and government funds were appropriated to bring in additional supplies from Hong Kong as well as to begin research studies on the culture of the local *Pangasius* catfish species. In the Philippines, aided in part by the United States under the large Philippine Rehabilitation Program, the Bureau of Fisheries developed private farms for both fish and oysters, and built many demonstration units for increasing the production of milkfish culture by converting mangrove swamps into fishponds.

New projects were not only confined to Asia. In Africa, the Fisheries Department of Egypt began a program for the development of more farms around the coastal lakes of Mariout, Manzalleh, and Karound. Further south, the new Experimental Fish Culture Station in Sudan was built in 1953 at Gordon's Tree. New commercial farms were built in Northern Rhodesia, Zanzibar, and Kenya, and a demonstration farm for rice and fish culture got underway in Tanganyika. In Australasia, a new marine biological station was built at Dunwich in Queensland, Australia, for work on oysters and mullet, and took a first look at the potential of the local prawns. Two new centers were established in Netherlands New Guinea at Sarong and Hollandia, and fishponds built in the Territory of Papua and New Guinea at Lae and Angoram. Several countries in Central and South America followed suit, many with assistance from the United States and Great Britain. Old interests in restocking inland fisheries were renewed in Argentina, Brazil, Colombia, and Peru, and exploratory missions were sent throughout Mexico, British Guinea, and countries of the Caribbean.

The work of the many pioneers being practiced far away in colonial research centers and in native fishponds throughout the 1950s slowly began to be supported by work at home in Europe in research centers that had been temporarily abandoned in the postwar years of austerity. For the most part, the early research still focused on the traditional freshwater and brackish-water fishes. Austria, for example, completed two new research and production stations at Kreuzstein and Wallersee, and revitalized most of the eighty-nine salmonid breeding centers. Finland built new fish breeding stations in Lapland and in south Finland. Belgium and France concentrated on enhancement programs with the common freshwater sport fish, such as carp, roach, pike, and tench, in neglected canals and waterways that had first to be cleared of aquatic vegetation. They also reinstigated their small trout farming industries, and France began work on migratory salmonids in its larger rivers. Italy turned to its ancient eel industry and built new stations at Marina di Pisa and Capodimonte. Yugoslavia constructed many new ponds for the production of common carp, which had been introduced in the early twentieth century in areas around its three big rivers.

Fortunately, the general thrust back to fish culture was not fixed entirely on the traditional species, and the emphasis on fisheries research in Europe permitted a new group of young scientists to turn their attention to a range of new species, particularly marine fish, crustaceans, and mollusks. Furthermore, applications of fish culture techniques were not necessarily confined to the direct production of food. On Maui and Oahu, two of the Hawaiian Islands, the state and federal governments cooperated in a project to produce tilapia in large quantities to be used as live bait in the skipjack tuna fishery and to clear the vegetation in the irrigation canals of the sugar cane and pineapple plantations. After tilapia were raised in the Paia hatchery, a building on Maui converted from some stables for pack-horses that worked the plantations, they were grown out in old Hawaiian fishponds and in raceway tanks. Although large numbers of tilapia were produced at an economic price, they did not prove to be such an attractive bait for tuna after all, and the project was eventually stopped.

9.3   New scientific discoveries in Japan

In Asia in the twentieth century, the mantle of aquatic farming skills had passed from the Chinese to the Japanese. Traditionally an agricultural nation, but with limited space and resources, Japan began a program of modernization and security for the future through industrialization and expansionism. Domestic food production could only be increased through intensification, and heavy investments into research and development were made in every form of agriculture and fisheries throughout the 1920s and 1930s. The shortage of food in Japan following the Second World War accelerated the pace at which knowledge of marine species was gained during this period, and inspired the application of new information to crude aquaculture production.

Production techniques were continuously tested and changed in the field, and many of them succeeded. Success was due in no small part to the natural ingenuity of the Japanese people, the availability of cheap labor, and simple determination. The hanging-culture method for oysters, for example, had become more commercial with the range of rafts, racks, and longlines that in the 1920s replaced a three hundred-year-old pole and net system, partly because it could collect more seed. The industry was becoming quite valuable when the war

Figure 9.1 Japan; the modern Chitose hatchery. (Original facility in the center is now a museum.)

started, and after the war, it was renewed with vigor. Floating rafts and racks appeared in almost every coastal prefecture of the country, including northern Hokkaido, in the national effort to produce food in face of the postwar shortage. Later, the cultured shellfish became available for export to renew trade and to earn hard currency.

There were many important pioneers in Japan during the years on either side of the war. The early research work on the culture of the abalone by Saburo Murayama in the 1930s had to wait for development pioneered by T. Ino after the war, and for the scallop, the work of Isahaya and Kinoshita built on advances made by Yamamoto. One of the principal pioneers who worked both before and after the war was Motosaku Fujinaga. His studies on the valuable Japanese shrimp, begun in 1933, were interrupted for seven years, but once the war ended, he returned to his research. He greatly improved hatchery production through the use of deeper tanks and by the replacement of his usual larval feeds by brine shrimp. Soon he was appointed chief fisheries research scientist of the government, a position from which he was able to exploit his vision for farming shrimp around the Inland Sea of Japan. In the mid-1950s, he bought some disused salt pans around Seto Island and converted them into the first ponds for shrimp farming. From there, his technology, which would be refined further by his young protégés, was spread throughout the region.

One of the most remarkable postwar developments in Japan was the sudden explosion of the seaweed industry. The culture of laver, the red seaweed of genus *Porphyra*, was just becoming a viable private industry in the country at the turn of the century, but it seemed to be greatly dependent on luck. Seeding the traditional brushwood stakes, called *hibi*, or the new poles and lines each season was a natural process, but always constrained by the temperament of the spores. Although large numbers of spores appeared to be present in the bays in the spring, there was frequently a dearth of settlement in the autumn. In 1949, a British scientist, Kathleen Drew, opened the door to artificial propagation by solving the mystery of the summertime disappearance of the sexual carpospores. She found that they did not persist in suspension until autumn, but rather, they quickly bored into old oyster shells in the substrate. There they grew into a filamentous plant that for a long time had been identified by botanists as a different species, *Conchocelis rosa*. It was this "new plant" that later released asexual spores that settled and grew into the next generation of edible thalli of *Porphyra*. With Drew's discovery of the relevance of this conchocelis phase, some Japanese scientists took notice. For example, Soukichi Sagawa, a botany professor at Kyushuu University, believed her hypothesis, but his botanical society would not. Also convinced was his friend, Fuo Ohta at the Kumamoto Prefecture Fisheries Research Station. Soon, Sagawa and Ohta were able to control the process in tanks, using crushed oyster shells and substituting artificial substrates for the settlement of the spores. Immediately, they involved the local fishermen in Sumiyoshi, and together, they went into mass production.

Laver yields in Japan subsequently soared, with bays in coastal provinces from Chubi to Kumamoto covered in myriad long hemp nets (later synthetic fibers), each capable of carrying enormous loads of the edible seaweed, also called *nori*. Drew was honored for her contribution to the development of the seaweed industry in Japan. The fishermen collected enough money to build a statue to her, but before she could visit Japan to sit for the sculptor, she died in 1957 at the early age of fifty-three. Instead, the fishermen erected a memorial to the lady they called the "Savior of the Laver Fishermen," which was unveiled on April 14, 1963, in the Sumiyoshi Shintou Shrine in Uto City in Kumamoto Province. Each year on that particular day, the Sumiyoshi Fisheries Cooperation continues to celebrate the Drew Festival. Beneath the memorial stone, the fishermen buried her scientific papers on her studies, together with her university hood and gown that she used while attending the University of Manchester in England.

The influence of the Japanese from this period on future aquaculture development was remarkably significant. This was due not only to their scientific and technical abilities, but also to their physical occupation of parts of China, including the island of Formosa. Throughout their lengthy presence, the natural affinity of the Japanese for fish and shellfish was not overlooked. The governor-general of Formosa established and organized an infrastructure for fisheries, much as it existed in Japan, and constructed a fish culture station at Tainan in 1910 under the distinguished scientist Takeo Aoki. First building on the culture of milkfish, which had been introduced centuries before, the scientists at Tainan worked on a variety of species, including key marine species, such as the

(a)

(b)

Figure 9.2 Japan; the Drew Festival and memorial in the Sumiyoshi Shintou Shrine, Uto City: (a) Drew Festival ceremony; (b) Kathleen Drew's memorial draped with gifts from the laver fishermen. (Courtesy of the Sumiyoshi Fisheries Cooperation.)

yellowtail, sea bream, and abalone, as well as freshwater eels and tilapia introduced from Japanese territory in Indonesia. Most of their studies complemented
research and development going on in Japan at the same time. Beyond the Tainan
operation, the government built several other stations throughout Formosa, all
of which were retroceded to the Republic of China in 1945.

Following the formation of independent Taiwan in 1949, these stations continued to work on furthering the research and development initiated by the
Japanese under the Joint Commission on Rural Reconstruction. This was funded
by the Rockefeller Foundation from the United States. Under the direction of
Tung-Pai Chen, fish culture in Taiwan became an important economic industry.
It began with the simple culture of tilapia in rice paddies, but in time it would
be recognized for the efficient farming of milkfish and fattening of river eels for
the Japanese market. Chen and his young associates were subsequently successful with many developments in aquaculture research, particularly the control of
breeding and propagation of silver carp, grass carp, and gray mullet for the first
time, and later with the culture of species of marine shrimp.

9.4   Hatchery propagation of oysters and marine fish in Europe

With the breach of Hitler's Atlantic Wall in 1944, it took little time for the French
oystermen in the postwar years to rebuild their famous industry to its historic
levels. From the beaches of Normandy round to the great Bay of Arcachon,
they reimplemented the classical system of artificial collectors on which they had
relied since Professor Coste carried out his experiments in the Bay of St. Brieuc.
Fortunately for the French oystermen, the sheltered Gulf of Morbihan in the
southwest corner of Brittany was the most prolific region in the world for the
production of spat of the European flat oyster. There was enough not only for
the French, but also for export to Spain and the Netherlands.

The Netherlands had long been an independent producer of the European
oyster, although on occasion, the stocks would almost be wiped out if the winter
temperatures were low enough to freeze the waters of Zeeland for a prolonged
period. Spain, in contrast, was not—partly because the oyster (unlike the mussel) was not regarded as a traditional food by the older generations, and partly
because efforts to grow them had not been very successful. However, interest
was once more renewed in 1954 in Galicia by Buenaventura and Pedro Arte,
two scientists at the Fisheries Research Institute in Vigo, and later played out
by Antonio Figueras. For the next ten years, they coped with one difficulty after
another, not all of which were technical. However, their efforts were rewarded
when they began to raise the oysters in the water column, as did the Japanese,
rather than on the bottom substrate. Their techniques caught the attention of the
local fishermen. Even though reliable production of seed was an ongoing problem, there were large quantities of spat available and supplied from Brittany.
Soon the *rias* of Spain were filled with a ramshackle collection of hulls of boats
that had been stripped of their superstructure and fitted with great wooden wings
from which were suspended ropes and bags containing growing oysters. Further,

the fishermen began to apply the same technique for the production of mussels. The traditional beds of mussels were abandoned as all the concessionaires turned to off-bottom culture, but they replaced the old hulls with well-engineered floating platforms and outriggers. The results were prodigious, and by 1970, Spain was by far the largest producer of mussels in the world—over 300 thousand tonnes annually. The farmers also had a reliable production of oysters.

For France, the lack of oyster spat had never been a problem. Not so in the British Isles, where in the 1930s, the accidental introduction of the parasitic slipper limpet and American tingle on oysters brought from the United States had almost wiped out the ancient beds of native oysters that had once fed the Roman conquerors. Research by the British government to reestablish the shellfish industry was curtailed by the Second World War. But throughout the late 1950s, as part of its general program on marine biological research, it supported two important applied research projects. One was on the propagation of the European flat oyster under Peter Walne at the Ministry of Fisheries' laboratory at Conway, and the other was on the breeding and propagation of marine flatfish under James Shelbourne at the Lowestoft Laboratory. In 1961, the British White Fish Authority, a quasi-government organization that was operated jointly with the private fishing industry, instigated projects to carry the promising research results of these two individuals through to commercial implementation.

The White Fish Authority constructed two experimental hatcheries, both of which had links to the past. An oyster hatchery was built at the old mussel depuration plant at Conway in North Wales. The plant had been constructed by the municipality in 1914 for the benefit of local mussel collectors. At the end of the First World War, the operation and site was taken over by the Board of Agriculture and Fisheries, which also added a laboratory to monitor local pollution and contamination of the mussel and oyster beds. The dream of operating a pilot-scale shellfish hatchery was finally made possible with the increasing reliability of Walne's techniques for rearing oysters. Building on the discoveries by H.A. Cole in the 1930s, which demonstrated that it was the nanoplankton that were vital to successful growth and attachment of oyster larvae, Walne developed an efficient system for producing the most suitable of these phytoplankters in very large quantities and overlapping their delivery to suit the diverse development of each larval shellfish population. Although shellfish were not supposedly in the mandate of the White Fish Authority, with the cooperation of the Ministry of Fisheries and the Shellfish Board, an agreement was reached.

The White Fish Authority's hatchery for marine flatfish was constructed at Port Erin in the Isle of Man. The Marine Biological Station was at that time a field center of the University of Liverpool, but it had been built originally in 1892 as a laboratory for the Lancashire and Western Fisheries Commission for its fisheries enhancement and transplantation program. Therefore, the laboratory was provided with some unusually large and deep holding tanks for cod and other adult marine fish. With the cooperation of the university, these facilities were called back into commission by the White Fish Authority in 1962.

Figure 9.3 Isle of Man, United Kingdom, 1964; marine laboratory containing White Fish Authority hatchery for flatfish built at Port Erin.

The successful operation and yields from the first true marine flatfish hatchery soon enabled follow-up demonstrations to be made at field stations. In 1964, a site was constructed at Ardtoe in the Western Highlands of Scotland for production of plaice in an enclosed arm of a sea loch. This was followed in 1966 by a site at Largs in Ayrshire, using the heated effluent from the Hunterston nuclear electrical generating station in an attempt to accelerate growth rates of Dover sole. Although the farming of flatfish at both Ardtoe and Hunterston was equally successful, production was not economical. Consequently, attention turned to the more valuable species, such as turbot and eventually to halibut, both of which proved more difficult to propagate.

Nonetheless, there were several benefits from these capital investments made by the White Fish Authority at that time. In addition to being sites for propagating and accelerating the growth of other marine species, such as abalone, Ardtoe and Hunterston were put on the itineraries of many international fisheries scientists, eager to observe farming the sea at first hand. At Hunterston, there were large delegations from Russia and Japan. The delegation from Russia in 1968 was most unusual, because it was made up of both fisheries scientists and nuclear physicists. There was also a continuous stream of individual biologists from all over the world, and many there were many national scientists, as well. One regular local visitor was Ron Roberts, a veterinarian at nearby Stirling University in Scotland on the look-out for new fish disease topics for his postgraduate students. The fish farm sites were fertile hunting grounds for research projects, and his interest in the new field grew. Subsequently, with funding and political awareness of the fisheries officers in the Overseas Development Administration some ten years later, his small program led to the creation of an aquaculture complex at Stirling University, which would support many bilateral aid projects in aquaculture development in the years to come.

The benefits of the White Fish Authority's work also fell on the other side of the English channel, where the French and Italians were interested in any new

Figure 9.4 Scotland, 1966; experimental flatfish farm site built by the White Fish Authority at Ardtoe.

Figure 9.5 Scotland, 1966; Hunterston Nuclear Generating Station in Scotland—the site where the White Fish Authority reared flatfish in heated discharge waters (background is Great Cumbrae Island, where the Millport marine station was opened in 1897). (Courtesy BNFL Magnox Electric plc, Scotland.)

activity that would bring more seafood to their tables. Scientists of both nations began to apply Shelbourne's technology not only to flatfish, but also to other marine species that were popular in their traditional fresh-fish markets. In the government laboratory at Brest, Michel Girin started to work on the bar or loup (the European sea bass) as well as the Dover sole and turbot. In the Languedoc region, Gilbert Barnabé also began with the bar and the daurade (gilthead bream) for farming in the *Lagunaire de Sète*, while Gino Ravagnan and Pietro Ghittino started to do the same for reinvigorating traditional fish culture in the *valli* around the coasts of Italy. Their applications were unquestionably successful, and farming of sea bass and gilthead bream soon became significant enterprises throughout the Mediterranean region, particularly in Greece some decades later.

9.5　The ubiquitous brine shrimp

There was one thing in common with Motosaku Hudinaga's (Fujinaga) sudden production of some fifteen million twenty-day-old marine shrimp postlarvae in his hatchery in Japan after thirty years of work, and Jim Shelbourne's successful production of one million juvenile flatfish at the White Fish Authority hatchery in the Isle of Man in Europe. Their mass propagation technology was made possible at that time only with the help of the small, pink nauplius of the brine shrimp, *Artemia*. This was the edible live food that was necessary for the survival of all their larvae (see Section 7.5). Without it, their attempts to advance the dream of farming the sea would have been impossible.

Consequently, in the mid-1960s, the encapsulated cysts of the brine shrimp were in great demand by the growing number of marine centers developing production-scale technologies and were suddenly wanted in unheard of quantities. The early processors were the commercial salt companies around the Great Salt Lake in Utah, which had begun to exploit the natural production of brine shrimp cysts in the evaporating salt ponds. They supplied small quantities to dealers in the growing tropical fish and aquarium trade, who repackaged them in small tins weighing a few ounces. Suddenly, they were deluged with orders for one hundred or two hundred kilograms. The processors had to scale up their production immediately and find new resources. Fortunately, these were to be found in the natural salterns of California, particularly in San Francisco Bay and down the length of the coast.

The reliable availability of large quantities of instant live food made it possible to scale up successful propagation of marine fish and shellfish in laboratories to production on a commercial hatchery scale. And the cost was low. Although the price of vacuum-packed cysts in the mid-1960s was well over $55 per kilogram, and most hatcheries wanted supplies in hand for two or three years ahead, the convenience of instant food in tins with unlimited shelf-life was far more economical than any investment in staff and the traditional wet laboratory techniques for live food production.

Brine shrimp cysts would become increasingly important for every marine culturist for the next thirty years—so much so that Guido Persoone and Patrick

Sorgeloos from the Ghent State University in Belgium went out on a limb to persuade the Ecology Department not only to prioritize research on *Artemia* but to establish the international Artemia Reference Center. Since 1985, the new facility has been an independent research arm of the Faculty of Agricultural Sciences, with a large complement of research and teaching staff. The Artemia Center made Sorgeloos the international expert on brine shrimp, and he was requested to search for resources all over the world. Both he and Persoone would go on to play important roles in the development of both the European Aquaculture Society and the World Aquaculture Society.

The high value of brine shrimp cysts also attracted considerable investment capital into some high-risk schemes to develop *Artemia* farming. Perhaps the most risky was in the remote Gilbert Islands (Kiribati), in the middle of the Pacific Ocean. The plan, supported by the Gilbert Island Development Authority, called for the integration of several lagoons within the giant atoll of Christmas Island. The climate of Christmas Island was noted for its enormous evaporation rate, and the idea was that by controlling sea-water flow mechanically, the salinities within the lagoons could be regulated, thus creating the right environment for the brine shrimp to produce encysted eggs. On a pilot scale, the system worked quite well, and the product was canned and sold. However, the full commercial venture was never put to the test, because the financial resources, according to some of the Gilbertese, were "personally redirected."

9.6 Fish and shellfish farming in North America

The first totally new fish farming business to make its mark was the catfish industry in the United States. Although the U.S. Department of Agriculture had promoted fishponds as a natural adjunct of farming and ranching in many of the rural states during the Depression and through the years of austerity in the 1940s and early 1950s, the innumerable acres available for rearing fish did little to contribute to national production, beyond meeting some very local demand. The most economical use of the ponds was raising baitfish for recreational purposes. However, with interest in fish culture intensifying once again in the late 1950s, and favorable legislation by the U.S. Congress, the catfish industry began to take off. It was helped on the farm by the successful propagation of channel catfish and blue catfish, which replaced reliance on the traditional buffalo fish, and the use of supplemental pellet feeds, which quadrupled the old yields from the ponds. Efficiency was also helped by the construction of purpose-built ponds rather than making do with the irregular ponds that were mostly shaped by drainage schemes.

The willingness for farmers to invest in well-laid-out farms in the 1960s was greatly encouraged by parallel investment in research and development at federal institutions in Arkansas and Alabama. Two of the pioneers in research on spawning the channel catfish through hypophysation in 1957 were Howard Clemens and Kermit Sneed, and their successful results paved the way for the construction of the national fish farming experimental stations at Stuttgart in Arkansas and

Marion in Alabama. With the popularity of farming catfish widening almost daily, the research and development capacity of the federal government was increased in 1965 by merging these two facilities with other regional laboratories to create the U.S. Warmwater Fish Cultural Laboratories. The national stations at Marion and Stuttgart were both carefully directed by Harry Dupree, who not only managed their research and development work for over twenty-five years, but also made his own significant scientific contributions to the catfish industry in fish nutrition and feed formulation.

Other institutions and individuals also played a part. After pointing the way for the industry with artificial feed formulations, Homer Swingle found it relatively easy to get funding to build up a nucleus of fisheries scientists and technologists at Auburn University, who would keep the institution in the forefront of catfish and tilapia culture for the next two decades. Among them was Claude Boyd, who would become the leading international authority on fishpond management and productivity; Tom Lovell, who would be a leading fish nutritionist and play a critical role in the production of catfish with the right taste for the processors; and Len Lovshin, who would help transfer Auburn's catfish and tilapia technology to the world's developing countries.

With all this attention, the catfish farming in the United States expanded into Mississippi, Louisiana, Missouri, and across to Texas. Subsequently, it would be found in another ten states, as far north as Illinois and as far west as California. Within fifteen years of the initiation of the federal program, some 25 thousand hectares of water surface were devoted to catfish farming in the country, and national production had reached an amazing 25 thousand tonnes. By the end of the century, it was well over 200 thousand tonnes.

With the organization of the catfish industry into structured farms producing harvests for human consumption in the 1960s, most operators of the older ponds lost an important part of their income. Consequently, many were forced to resort back to the production of goldfish and minnows for bait. In the southern states around the Mississippi Delta, many of the farms had a feral stock of crawfish, which had an attractive specialty market that was mostly small and local. The opportunity for expanding crawfish production appealed to Jim Avault and Larry Le Bretonne at the Louisiana State University, and through their pioneering and practical research in the field, they began to test and introduce a variety of culture and management practices to the farmers. Soon they were joined by Jay Huner, who through his personal enthusiasm and flare for publishing articles about crawfish wherever he could, would spread their crawfish farming technology all over the world. By the end of the 1970s, the small Louisiana team had helped to make the crawfish an individual species for culture in the United States, as well as one incidental to other fish farming operations. Aided by their enterprise, national crawfish farmers would be producing about twenty-five thousand tonnes of the popular crustacean every year.

Far away from embryonic catfish industry of the humid states of the Mississippi Delta, in the cold Atlantic waters of Long Island Sound on the northeastern coast of the United States, an older industry was looking for help. In a matter of fifty years, the industry for the production of the American cupped oysters had plummeted from four million bushels a year to a miserable fifty thousand bushels. By 1955, the industry was on its knees. Companies unable to withstand the poor spat-fall and high mortality losses year after year pulled out of the business.

Revival started around one man, Victor Loosanoff. Loosanoff was born in Russia and immigrated to the United States as a young man. His interest in natural history led to his work on mollusks, mainly oysters, which he often shared with another equally transposed countryman called Paul Galtsoff, who would became famous for his landmark work called *The American Oyster*. In 1931, Loosanoff was assigned by the U.S. Bureau of Commercial Fisheries to study the problems of the American oyster industry in Connecticut, and the government constructed a special laboratory for him at Milford on the shores of Long Island Sound. The work began slowly. Taking his cue from the well-organized oyster industry in France, Loosanoff and his young staff first began to assist the local farmers on the settling season by providing information about plankton analyses and settling forecasts. Their routine work helped, but any gains were quickly nullified by the years of the Second World War.

Fortunately, in the postwar years that brought the industry nearly to its death, things changed dramatically. With Loosanoff as the hands-on director, the research and development carried out at the Milford Laboratory suddenly became the focus of all major advances for shellfish in North America. The most important discovery was the technique for production of cultchless seed for both American cupped oysters and clams. The problem of oyster culture had always been rearing through the settling stage. Hatching the American oyster indoors had been achieved as early as 1879 by William Brooks, but it took another forty years before William Wells managed to carry larvae through to the setting stage for the first time. Traditionally, old shells or lime-covered tiles were used for attracting spat in the settling season, but the goal of many culturists was to carry spat production through the settlement stage in a hatchery. This would bypass the laborious and wasteful process of chipping each spat free from its substrate by hand. By modifying Wells' technique and using artificial methods to induce spawning, Loosanoff and his colleague H.C. Davis had the break they were looking for. However, the technique still used very clean and sterile oyster shell as the cultch for the settling larvae, which added a dimension of considerable labor to such an operation in the hatcheries that were built. But it was not for long, because in a matter of a few years, Loosanoff together with an oyster grower from California called Bill Budge developed the technology for cultchless seed production. Thus young oysters and clams came out of the hatchery as unique individuals and were ready for transfer directly to fine-mesh screen trays for further grow-out at the hatchery or on the nursery grounds. In addition to developing methods for spawning bivalves almost all year-round, Loosanoff and his colleagues at Milford went on to develop strains of oysters for fast growth. Just like Professor Coste, who a century earlier saved the oyster industry of France with his lime-covered ceramic tiles for settlement, Loosanoff and his colleagues at the Milford Laboratory salvaged the shellfish industry in North America. Within fifteen years, the growers in Long Island Sound were producing numbers back to their historical peaks. Today, many Milford techniques are still used worldwide by the shellfish aquaculture industry.

9.7   The farming of salmonids in salt water

The farming of salmonids in salt water began virtually in parallel on either side of the Atlantic in the late 1950s. In the United States, extensive experiments were made to raise Pacific salmon in saltwater lagoons around Puget Sound in Washington State to strengthen local fisheries. A number of lagoons were enclosed, and the young salmon were left to forage for themselves. The experiments were not successful because of the shortage of food, high summer temperatures, oxygen depletion, algal blooms, and disease. Under the inspiration and direction of Tim Joyner at the National Marine Fisheries Service (NMFS) laboratory in Seattle, an alternative was proposed. The "Brown Bear," a small reconditioned coaster from Alaska, was converted to a floating hatchery ship. The idea was to raise smolts in salt water and then hold them back before release in various parts of Puget Sound, where they would remain. A base for the vessel was found in the grounds of a naval fuel depot at Manchester, across Rich Passage from Bainbridge Island. In addition to the tanks on the deck of the vessel, a flimsy complex of floating cages was constructed around the side in which could be held more smolts. Later, it was expanded to hold broodstock.

Although the principal goals of the work by NMFS were to continue to advance salmon culture technology and to increase production of the Columbia

Figure 9.6 United States, 1971; floating hatchery, MV Brown Bear (at Manchester pier for Pacific and Atlantic salmon).

River system, the relocation to Manchester Field Station in 1969 broadened the emphasis to include culture in the marine environment. In their floating net-pens, the young coho salmon, which were naturally more coastal in their migratory habits, did particularly well. They grew rapidly after smoltification to twelve-inch size and three-quarter-pound weight. The obvious deduction by the NMFS team was that these "pan-sized" fish could in fact be marketed directly without release into the fisheries at all. Consequently, by this fortuitous twist, Joyner, together with his colleagues at Manchester, Tony Novotny, and Connie Mahnken, and with Bill McNeil working independently at Oregon State University, initiated coastal farming for coho salmon in the Pacific Northwest.

On the other side of the Atlantic, the Norwegians were anticipating the growing problem of a declining fishing industry, which was the principal economy of most of the nation's coastal communities. In the early 1960s, the government began to invest in research and development, and financed early investors through its Regional Development Fund. However, this time they concentrated on the high-priced Atlantic salmon instead of Atlantic cod and other similar marine species that had been the subject of their programs almost a century before.

The challenge of rearing anadromous fish in captivity had been taken up long before. Since the mid-1930s, scientists at the Swedish Salmon Research Institute at Sundsvall, led by Börje Carlin, had successfully reared young Atlantic salmon through to parr, and at the end of the war began to enhance the fishery in the Baltic Sea by releasing large numbers of smolts every year. Because the fish returned of their own accord as adults, the practice was soon described as "ranching." The activity attracted much attention, and new salmon research stations were built in Finland, Iceland, the Faroe Islands, Scotland, and Ireland to propagate and release smolts. Later, Carlin's Institute at Sandsvall was relocated further south to Alvkarleby, and nearer to Stockholm.

The pioneers of commercial farming of Atlantic salmon were the two Vik brothers from Sykkylven, and Ivar Heggen, who worked at Vike-oyra in Norway. For some years, Karstein and Olav Vik had been interested in rainbow trout, a species readily available in freshwater fish farms in Denmark since its introduction half a century before. They had managed with great care to introduce rainbow trout into full sea water, where they grew very well on dry-feed pellets. In 1955, the men then turned their attention to raising Atlantic salmon. After some years of trial and error, they began their experiments to rear smolts in full sea water. They launched their first floating cage, a simple wooden box, in 1959. The trials were a success, and forty fish reached maturity by 1962. These fish were reintroduced to fresh water and spawned, thus completing a life-cycle entirely in captivity.

The pioneering work of the Vik brothers attracted the attention of the traditional freshwater rainbow trout growers, and there were many other attempts made in other coastal areas of Norway. However, progress was slow, primarily because of the difficulty of the transition from fresh water to salt water, and that of producing the large quantities of smolts necessary to make farmed production economically viable. One of the first companies to undertake production on a large scale was Mowi A/S. Thor Mowinckel built four sites near Bergen: two

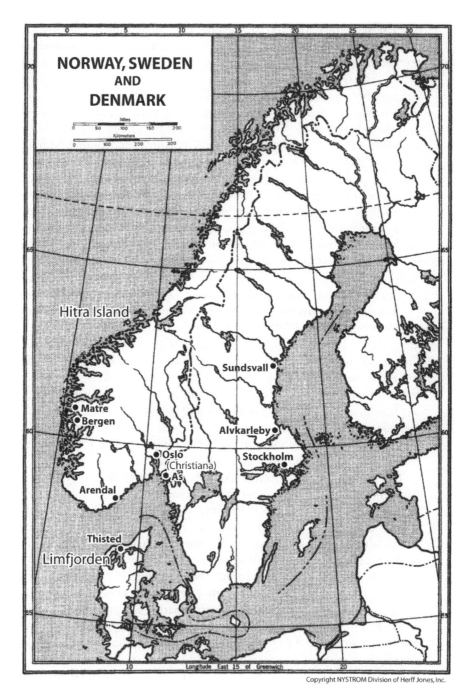

Figure 9.7 Norway, Sweden, and Denmark. (Adapted from basic map, copyright NYSTROM Division of Herff Jones, Inc.)

Figure 9.8 Norway, 1978–1980: (a) Averøy, sea cages, 1980; (b) Averøy, advanced sea cages; (c) Averøy, the first automatic feeder, 1978; (d) Averøy, pens for feeding trials, 1980.

hatcheries for the production of smolts only, and two coastal enclosures for grow-out. This separation of the two activities essentially established the operational system around which the industry developed throughout the late 1960s and into the 1970s.

The success of commercial salmon farming in Norway was spectacular, and records for smolt production and harvest were broken year after year. By the end of the decade, national production was over 4 thousand tonnes. Twenty years later, it would stand at well over 300 thousand tonnes.

Across the other side of the North Sea in Scotland, the first attempts to farm salmon were made by Marine Harvest Ltd., a subsidiary of Unilever Corporation, the giant multinational company. As Lever Brothers, the company was no stranger to investing in the Outer Hebrides. At the end of the First World War, the first Lord Leverhulme, the founder and philanthropist, bought Lews Castle and tried to bring economic development and social growth to the islands, just as he had done elsewhere in England. This effort was cut short by his death in 1925. Marine Harvest's interest began in 1966 with rainbow trout in salt water at Lochailort on the west coast of Scotland, not far from the White Fish Authority's marine fish site at Ardtoe. However, with the publication of the results of the Vik brothers and the great interest in Norway, the company quickly added Atlantic salmon to its research and development program. Soon it was joined by other companies, particularly the Highland Trout Company Ltd., which had established a base at Otter Ferry, lower down the west coast.

Most of these early enterprises tried both coastal enclosures and floating cages. Several early constructions met with structural disaster, and the operations suffered loss of the captive fish, which were the easy prey of marine mammals and birds. One remedy was offered in Norway during the late 1960s, when two brothers, Sivert and Ove Gröntvedt, designed and constructed a large octagonal cage with a strong fixed collar that also acted as a servicing platform. The cage could be covered to keep out predators and could be nested together for easier servicing. They tested their ideas in the waters of Laksevika, on Hitra Island west of Trøndheim, and their designs became the prototype for the majority of cage farms that were soon to follow.

9.8   Freshwater prawns and gray mullet in the Pacific

In the United States, funding for any new technology has always been relatively plentiful for both public and private research. Consequently, beginning in the 1960s, many biologists and marine scientists all over the country began independent research on the culture of species they thought potentially useful. Their interest had been sparked by two events: creation of two new government-related bodies, very actively engaged with marine affairs, including aquaculture.

The first was the creation of the President's Scientific Advisory Committee by Lyndon Johnson when he stepped up to the position in 1963. The committee was very actively led by Vice President Hubert Humphrey for five years. Its objective was to explore the resources of the sea through an intense national program of oceanographic and marine research, somewhat similar to the space program a decade before.

The operational leadership for the task was given to Chairman Julius Stratton of the Ford Foundation. As part of the fact-finding work on all aspects of the sea by a host of scientists in various special committees, John Bardach and John Ryther were dispatched worldwide to study and report on the status of marine farming. In spite of their unfamiliarity with this totally new field, it was an opportunity they could not turn down, and subsequently they used their experiences to write one of the first general books on aquaculture. This bestseller, simply called *Aquaculture*, introduced the field to much of the developed world for the first time.

The Stratton Commission's Report on Marine Science, Engineering, and Resources was published in 1969. It was the nation's blueprint for what the media nicknamed the Blue Revolution. Aquaculture was singled out by the following words: "Major new efforts directed towards [*sic*] the understanding of the reproduction, growth, and development of potentially exploitable marine organisms should be undertaken to provide the base of understanding and technology necessary to make the products of aquaculture more available."

The second event that focused the attention of American marine biologists on aquaculture had been set in motion before the publication of Stratton's report. Stratton's strategy to achieve the objectives was to restructure the principal government agencies involved in marine affairs and to create a new agency in the National Oceanic and Atmospheric Administration (NOAA). This was the Office of Sea Grant. Conceived to be like the old Land Grant Program, which had been responsible for creating the nation's highly productive and prosperous agricultural sector, Sea Grant was designed to be the conduit for government funds to be used in association with private capital and the resources of research institutions. Soon, however, Sea Grant would become one of the few real testimonies of the Stratton report, as the general euphoria for a Blue Revolution was slowly suppressed by the disinterest of Richard Nixon's administration that took over in 1969.

Two of the first projects supported by Sea Grant were in Hawaii, and both would emerge to make important contributions to the global development of aquaculture. The fact that these and other projects were funded so quickly was not surprising. Many scientists, businessmen, and politicians from the small marine state of Hawaii were members of Stratton's various committees, and they were ready to capitalize on the government's interest in marine science and technology. The director of Sea Grant was Hal Goodwin, who as a former journalist and the first incumbent of the office, gave himself the latitude to follow his trained nose for hunches. Consequently, he was prepared to back good ideas outlined on the backs of envelopes, together with their rough budgets, and for over a decade, his support of research and development in aquaculture in the country was a *tour de force*.

The first project concerned freshwater prawn farming. In 1965, the State Department of Land and Natural Resources imported thirty-six giant freshwater prawns from Malaysia, and under the watchful eye of Takuji Fujimura, began a program for propagation and subsequent farming. Fujimura used the basic breeding and rearing techniques first developed by Shai Wan Ling, the FAO Regional Fisheries Officer in Bangkok some years before. In an attempt to discover anything that would keep his dying prawn larvae alive, Ling had dropped soy sauce into the freshwater hatching tanks, and that touch of *salt* had proved to be all that was required.

Fujimura and his team, funded by the Hawaiian government and aided by the Office of Sea Grant, soon solved the local problems of feeding and harvesting, and the operation was scaled up to demonstrate its commercial potential. By the early 1970s, the foundation of a freshwater prawn industry for Hawaii was established. The state built a large hatchery on Sand Island in Honolulu Harbor and began to supply interested farmers with free seed and advice. By the end of the 1970s, there were a dozen or more farms spread throughout the Hawaiian Island chain active in freshwater prawn cultivation, and production was rapidly absorbed by the local market in Honolulu and the hotels. Fujimura became the recognized expert in the field and assisted many small industries to develop all over the world, including introducing freshwater prawn farming back into Asia.

The second key Sea Grant project was taking place on the other side of Honolulu at the Oceanic Institute, where Ziad Shehadeh and Ching-Ming Kuo were further refining the Taiwanese techniques for the controlled spawning of the gray mullet. Shehadeh had spent a year working with I-Chiu Liao and his colleagues at Tungkang, where spawning had been induced for the first time by Yuan Tang in 1964. Much of the fish culture research in Taiwan on gray mullet, milkfish, and eels was part of the long-running Program for Rural Reconstruction and Development, funded by the Rockefeller Foundation. The Office of Sea Grant began to fund a parallel project to complement the research support of the Rockefeller Foundation for the Oceanic Institute's program on brackish-water fish culture.

Regrettably, neither of these two pioneering projects into aquaculture by the Office of Sea Grant had lasting benefits for the aquaculture industry in the United States, primarily because farm production was never economical. The early optimistic forecasts for profitable farming of freshwater prawns in Hawaii proved to be marginal at best, and investment interest was lost when marine shrimp technology became more available. Similarly, the work of Liao, Shehadeh, and Kuo (and others such as Hiralal Chaudhuri at Cuttack in India, and Abraham Yashouv at Dor in Israel) on the breeding and propagation of gray mullet never resulted in either subsistence farming in the Pacific Islands or commercial farming in Hawaii, as was intended. This was largely because the mullet was relatively unimportant in commercial markets, and therefore, hatchery and farm production was always contending with poor consumer perception and the resulting low prices. However, their fundamental research into the use of gonadotropins and other hormones for induced breeding, and the hatchery procedures they established, forged the way for breeding and propagating many other species of marine fish, such as milkfish, grouper, and sea bass.

Bibliography

Bardach, J., Ryther, J., and McLarney, W.O. (1974) *Aquaculture: The Farming and Husbandry of Freshwater and Marine Organisms.* John Wiley & Sons, Inc., New York.

Galtsoff, P.S. (1964) The American oyster, *Crassostrea virginica* Gmelin. *Fishery Bulletin of the Fish and Wildlife Service,* Vol. 64. Government Printing Office, Washington, D.C.

Hickling, C.F. (1962) *Fish Culture.* Faber and Faber, London.

Huet, M. (1952) *Traité de Pisciculture (Textbook of Fish Culture).* Brussels [in French].

Huet, M. (1970) *Traité de Pisciculture (Textbook of Fish Culture),* 4th edn. Editions Ch. de Wyngaert, Brussels [in French].

Schaeperclaus, W. (1933) *Textbook of Pond Culture: Rearing and Keeping of Carp, Trout and Allied Fishes.* Paul Parley, Berlin. [in German, translated to English in 1948].

[Stratton] Commission on Marine Science, Engineering and Resources (1969) *Our Nation and the Sea: A Plan for National Action.* U.S. Government Printing Office, Washington, D.C.

Chapter 10

Uncontrolled Expansion (1965–1975)

Abstract

From 1965 to 1975, uncontrolled aquaculture expansion derived from North America and European oyster and clam hatcheries. Marine flatfish were farmed in Scottish lochs. Edible seaweed, marine shrimp, and yellowtail were cultured in the Inland Sea of Japan, and freshwater prawns in Honolulu; eels were grown in Taiwan raceways. Growth also resulted because inexpensive plastic ponds replaced earth ponds, fiberglass tanks replaced concrete, and new net-pens and tank plumbing appeared. Florida and Texas marine shrimp were hatchery reared, as were net-pen salmon at National Oceanic and Atmospheric Agency's operation in the Pacific Northwest. New technology, financed by international corporate holdings, spread throughout the Americas. Private hatcheries, farms, and ocean ranching followed in the Americas, Alaska, and Norway, along with large government salmonid hatchery programs in Japan. Many large and small businesses nurtured burgeoning aquaculture in the 1970s; few would be playing the game ten years later.

10.1 Introduction

Throughout the 1960s, a number of very different farming activities in very different corners of the world captured the interest of scientists and their governments alike. The principal activities were as follows: the use of oyster and clam hatcheries in North America and Europe to revitalize flagging shellfish industries; the rearing of marine flatfish in an enclosed sea-loch and at a nuclear power generating station in Scotland; the increase in production of edible

The History of Aquaculture. By C. E. Nash. Published 2011 by Blackwell Publishing Ltd.

seaweed, marine shrimp, and yellowtail in the Inland Sea of Japan; raising of freshwater prawns in Honolulu Harbor in Hawaii; and the growing out of eels in raceways fed by the bubbling aquifers of central Taiwan. After lying essentially dormant for more than half a century, the age-old idea of productive farming of aquatic animals and plants, and particularly of farming the sea, was given new life by these five pioneering efforts. Furthermore, the excitement rubbed off on the seemingly passive efforts that had been given to traditional practices of freshwater fish farming for decades.

As information about tomorrow's world was broadcast around the globe, helped by the medium of television by then in almost every home, the late 1960s and early 1970s witnessed an explosion of effort by groups of individuals who believed that the time for modern aquaculture had arrived. Yet, there were few who could possibly call themselves aquaculture specialists, and their number was largely made up of biologists, agriculturists, marine scientists, a handful of engineers, and several entrepreneurs experienced in business. This was the motley collection of people who set about to provide an alternative source of seafood for the world markets. In Norway and England, they focused their attention on marine production, working with marine fish, Atlantic salmon, and rainbow trout. In North America, they reinvigorated their work on Pacific salmon, and in the southern states, on the farming of catfish. Israel, in its continuing search for food self-sufficiency, stepped up work on carp and tilapia production through integrated farming. Hungarians worked on carps and adopted Chinese methods for integrated farming with ducks. In Japan and Taiwan, they intensified their work on the production of marine shrimps, and there were parallel efforts in the United States, Panama, Ecuador, Thailand, the Philippines, Indonesia, and England. In the Philippines and Indonesia, they set about to improve milkfish production. The Thais picked up on the work in Hawaii and pioneered further development of freshwater prawns. The Japanese expanded their work on abalone, sea bream, mackerels, and edible marine algae, while the Taiwanese improved the production of milkfish and included gray mullet. In Spain and the Netherlands, they began to produce prodigious quantities of mollusks suspended from floating rafts. The French worked on sea bass, flatfish, and oysters, and the Italians on sea bass and sea bream. In other European countries, the Danes, English, and French revitalized their flagging trout industries, and in Germany, it was the eel industry that benefited. Almost none of this decade of intensive effort would have been possible without the development of other complementary fields and technologies.

10.2   Help from other new technologies

The new age of aquaculture development that burst onto the world stage in the 1960s, like many other old and new industries, was only made possible by the parallel advances in other technologies. For aquaculture, the most important of these by far was plastics. Plastics technology, first with polyethylene and then polyvinylchloride (PVC), revolutionized the design and construction of all the life-support systems and tank complexes in every wet laboratory in every government institution exploring the new field of aquaculture. Subsequently, it would enable the construction of relatively cheap and clean commercial hatcheries for the species under culture. Plastics technology made it possible to replace the traditional thick iron pipes and valves, or the fragile glass pipes characteristic of the late nineteenth and early twentieth century aquariums and hatcheries, with light, inert, noncorrosive pipes and fittings. Furthermore, entire water and compressed air systems could be assembled very easily and quickly, without the expertise and equipment of skilled fabricators.

Plastic piping was extremely versatile. It was used not only for water and air lines, but also for construction. It was frequently assembled in a simple framework to support floating plastic sheet or fabric tanks for production trials. Subsequently, large-bore plastic pipe was used to make the floating collars from which some of the first net-pens were suspended. This, however, proved to be a dead end. It was a costly form of construction in terms of the price of the raw materials and of lost fish. The PVC pipes cracked in very cold water and sank, letting valuable fish spill out from the top. The cure was to inject the pipes with plastic foam, and then it became clear that the cheapest alternative was simply to fabricate foam-filled floats. This enabled the construction of nests of floating cages and rafts, complete with walkways and landing platforms. Such cages became popular, because they were effective substitutes for costly excavated ponds that required pumped water and the purchase of expensive, flat land on which to construct them.

Plastics technologies also revolutionized the fabrication of hatchery tanks, which traditionally had been made of wood, or more recently, concrete asbestos. But the first plastic tanks were very expensive, and their potential market in an emerging aquaculture industry was pre-empted by advances in another key technological breakthrough: the development of fiberglass.

Many of the early fish hatcheries of the twentieth century followed the traditions of Victorian public aquariums, with large concrete or wooden tank structures that had plate-glass viewing windows that invariably leaked through the metal frames, producing unsightly and lethal rust. Fiberglass processing suddenly enabled the construction of lightweight tanks, which could be moved easily from place to place and reused many times. Tanks of any required size and shape could be fabricated completely from a single mold or from molded panels that could be bolted together. The insides of the tanks were smooth and easy to keep clean. No internal painting was necessary, and the tanks could be colored as necessary. The material also made it easy to configure a tank to allow for any sort of plumbing arrangement.

Another timely development gladly accepted by aquaculturists was the modern scientific equipment that began to appear on the market. Prior to the 1960s, much of the analysis for the most common water quality parameters was a slow, tedious business carried out by white-coated technicians in a chemistry laboratory. Data were not available in a hurry, even in an emergency situation. However, with the increasing interest in ecosystems and the emergence of environmental sciences, the scientific instrument catalogues were filled with a range of hand-held instruments to directly measure and monitor the principal water quality parameters, such as dissolved oxygen, salinity, and pH. Also, there were simple back-pocket colorimetric test kits, which could be used outdoors around the fishponds to spot-check levels of other important variables that might spell disaster if they got out of line.

In addition to the benefits of these new technologies for the growth of the aquaculture industry, two other important changes were underway. One concerned marine fish and shellfish nutrition, and the other, marine diseases.

Although pelleted feeds had been available to freshwater salmonid hatcheries and trout farms in North America and Europe for fifty years or more, the logistics of feeding large populations of marine fish and shellfish in captivity every day boggled the mind. For the pioneers, the principal food was simply other fish and shellfish that had been freshly chopped or minced. For established juveniles in the hatcheries, the food was fresh mussel meats. When fresh mussel became too time-consuming to prepare, the daily diet became boiled mussel. As the locally available resources of mussels were depleted, the diet was changed to freshly chopped white fish. This was preferred to herring, which was frequently more available, and available in quantity, but chopped herring left an oily film around the tanks.

It was clear that something more had to be done when the marine fish or shrimps were out of the hatchery. For research and development work, manual preparation of fresh fish diets was acceptable, but for grow-out in commercial conditions, where labor and time translated to money, it would not do. Fortunately, a number of the commercial feed producers were persuaded to focus on the problem.

In the general euphoria of the biologists, nothing seemed impossible. Government development funds poured in, on the promise of taxes to be paid on the large financial profits of wealthy investors in the new technology, or of wide-ranging social benefits of cheap protein for the rural poor. New collaborative organizations and professional societies were formed, and the field was called "aquaculture."

10.3   The globalization of marine shrimp farming

On the eve of the outbreak of the Second World War, Motosaku Fujinaga made an important discovery: he could finally keep alive the zoea of his beloved *kuruma* shrimp by feeding them a diatom called *Skeletonema*. This at last was the breakthrough that he and his colleagues deserved after six years of patient research in a remote laboratory on the southern island of Kyushu. The laboratory was a field station of the Hayatomo Fisheries Research Institute, a private enterprise belonging to Kyodo Gyogyo, later to become the internationally renowned Nippon Suisan Company. However, another twenty years would pass before Fujinaga and his young team, which included Hiroshi Kurata, Jiro Kittaka, and Kunihiko Shigueno, had put together a sufficient number of the pieces to construct a hatchery using large and elaborately designed tanks, and to begin modest industrial production of the *kuruma abi*, which means in Japanese, the "shape of a wheel" from the way the exoskeleton of the shrimp is arched.

In 1963, Fujinaga started farm production operations in thirteen hectares of ponds the company constructed in the river estuary at Aio, a small town on the edge of the Inland Sea on the southern tip of Honshu. But he was not alone. His work had been followed closely for years, and there were several others wanting to try their luck at farming shrimp. Between 1960 and 1964, ten small shrimp farms were started, clustered around the sheltered and warm waters of the Inland Sea. Nine years later, their number had doubled. Some of them started their private businesses by taking advantage of many disused salt-beds that dotted the coast and creating ponds inside of them.

Good money-making business opportunities in Asian countries do not remain a secret for very long, and soon Fujinaga and his team were sought after by entrepreneurs outside Japan, from places such as Taiwan, Thailand, Indonesia, and the Philippines. Furthermore, these countries had other varieties of large commercial shrimp, which meant that research and development was refocused on a range of new and interesting species for culture. Most of the species adapted easily to Fujinaga's tank technology, and at least temporarily, his hatchery design method became textbook. Later, it would all go out of the window.

For over twenty-five years, shrimp farming technology was developed entirely by the private sector in Japan. The government gave it no financial support, and none of the government research institutes carried out any work on the problems. However, this had all changed by the end of the 1960s, when the government intervened to raise shrimp for restocking the coastal fisheries. Fujinaga's technology ceased to be a private monopoly, and the government essentially distributed it freely around the region starting in 1971, when it agreed to help the Southeast Asian Development Center (SEAFDEC) construct a large training, research, and production facility for marine shrimp at Leganes on Panay Island in the Philippines. This opened the door for entrepreneurs in any of the five SEAFDEC member countries to take the technology and apply it as best they could. In the Philippines, one early pioneer was Domiciano Villaluz. Villaluz was a professor at Mindanao State University, and subsequently became the scientific director of the Aquaculture Center at Leganes. Seeing the opportunity placed first at his own feet, he soon built up a substantial private farm operation on Mindanao Island for raising the giant tiger shrimp, and quickly many of his associates followed suit.

The fact that Indonesia, the Philippines, and Thailand all rapidly passed Japan in terms of farming marine shrimp in less than a decade was no surprise. First, there were other indigenous Asian species of shrimp besides the giant tiger shrimp that were more appropriate for farming than the fastidious Japanese *kuruma*. Second, their tropical latitudes were more conducive to fast growth than even the southernmost parts of Japan. Finally, large coastal areas were already covered with ponds constructed from old salt pans for the culture of milkfish, and the rich organic substrate, called *lab-lab*, proved to be equally ideal for feeding young shrimp.

The use of milkfish ponds for raising marine shrimp, however, proved to be something of a social dilemma. For some centuries, the farming of milkfish in Southeast Asia provided a vital source of cheap protein for the poor rural

Figure 10.1 The Philippines, 1972; shrimp production research ponds: SEAFDEC Aquaculture Center at Leganes, the Philippines. (Behind is the site of the United States Agency for International Development/Auburn University milkfish research project; beyond are traditional milkfish ponds.)

communities. In theory, the discovery of a more valuable crop held out hope of an economic boost for these communities. In practice, it did not turn out that way. Most of the large tracts of coastal milkfish ponds were quickly purchased by private companies and wealthy individuals for immediate conversion to shrimp ponds.

The transformation of the coastal milkfish ponds was subsequently accelerated by massive marine shrimp development projects financed by the Asian Development Bank and the World Bank. The reasoning of these institutions was that the new wave of shrimp farms would be beneficial for the economy of the coastal regions as a whole, and that the village communities would find permanent employment on the farms. But it would take twenty years to recover the displaced milkfish production in both Indonesia and the Philippines, and it would take yet more bank loans to construct yet more ponds and canals in exploitable swamp lands. This time, however, further marine shrimp development would be at the expense of the mangroves.

In Taiwan, where suitable coastal areas were not available, the country's few productive milkfish ponds were important to the national policy of self-sufficiency. Consequently, at the Tungkang Marine Laboratory, I-Chiu Liao and his colleagues temporarily switched their pioneering research studies from the breeding and artificial propagation of the gray mullet to the production of the giant tiger fry and rearing the fry in polyculture with milkfish.

News of these advances spread to the rest of the world. With continuous access to its old colonies and overseas territories in the tropics, the French government started a program in 1972 to culture marine shrimp and freshwater prawns at its Pacific Oceanological Center in Tahiti. The program was coordinated by Alain Michel. Because he was strongly supported by liberal government finance, he put together an aquaculture team that at times numbered well over thirty biologists and engineers. With the rotation of scientists through the center, the program at Tahiti was responsible for making well over a hundred French national and other professional experts in all aspects of shrimp culture. Many of these people subsequently become shrimp farmers in countries all over the world. Others would become project managers and leaders in international development projects awarded to France-Aquaculture and other French consulting companies. In addition to upgrading the entire level of expertise in marine shrimp culture, the French center at Tahiti made one other global contribution to the growth of marine shrimp farming. By the end of the 1970s, the simplified hatchery technology developed by the French had superseded that originally developed by Fujinaga in Japan and would become the model for almost every marine shrimp hatchery built throughout Asia for two decades thereafter.

One interesting feature of the work in Tahiti was that few of the very many scientific publications emanating from the Oceanological Center were credited to individual members of the team. All the important papers for almost twenty years were published under the acronym of AQUACOP—the aquaculture team of the Centre Océanologique du Pacifique.

As the largest consumer of marine shrimp in the world, the United States could not be left out of the free-for-all in the 1960s. Like other countries with extensive coastal regions covered with natural tidal lagoons, the United States had a small supplementary domestic marine shrimp harvest through the management of these rich areas. For some years, thanks to the work of Bob Lunz in South Carolina throughout the 1950s, the owners of these productive lagoons had made a modest contribution to their annual incomes by the Asian practice of trapping the juvenile shrimp that entered with the tide, and feeding them minimally, as Lunz directed. Later, the shrimp were harvested when they tried to migrate back to deep water to spawn, and then were sold on the local markets. There were many such impoundments in the south of the country, stretching from South Carolina, around Florida, and along the Gulf Coast shoreline, but the yields were relatively small, because the lagoons could never be cleared of predators. However, with the news that juvenile marine shrimp could be raised in hatcheries, there was a clear opportunity to intensify the production operations and to stock larvae in tidal enclosures that could be totally controlled.

Harry Cook and a handful of scientists at the NMFS Gulf Coast Center at Galveston, Texas, had been following Fujinaga's results with interest, and they arranged through the government for him to come and work with them in 1963. Together, they successfully spawned and reared two indigenous species, the brown shrimp and white shrimp. This was enough to persuade the state of Louisiana to construct the first experimental shrimp farm on Grand Terre Island, Louisiana, in 1964. For expertise, Cook's group encouraged the involvement of the NMFS Tropical Biological Laboratory in Miami and the Miami Institute of Marine Sciences, where Jay Ewald had successfully closed the life cycle of the

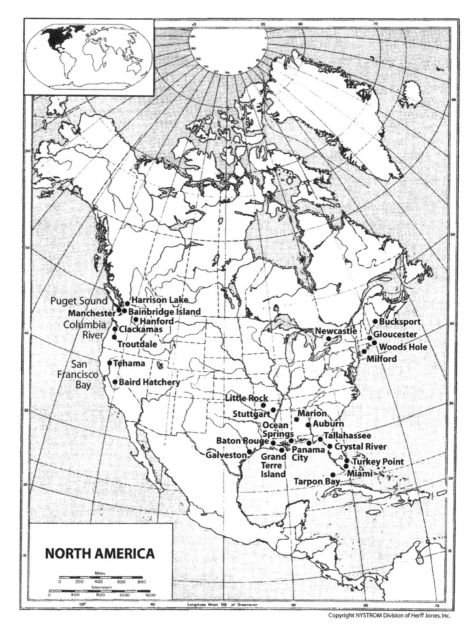

Figure 10.2 North America. (Adapted from basic map, copyright NYSTROM Division of Herff Jones, Inc.)

pink shrimp in 1963. Jerry Broom was appointed to manage operations, which he did until 1968.

The interest in this new opportunity for marine shrimp farming was whipped up by Harold Webber, an extraordinary entrepreneur with a profitable list of patents to his name in many different fields. Much to his credit, he had long considered that aquaculture was a business for the future and that much use could be made of the heated waters from electric generating plants. Consequently, he financed his own world tour in the mid-1960s to visit Japan to see what was being done there with marine shrimp. He then went through the United Kingdom to visit all the sites where the White Fish Authority was pioneering fish culture, especially at Hunterston in Ayrshire, where the site was supplied by thermal waters resulting from electrical generation.

A familiar figure in many executive boardrooms of large American corporations, Webber persuaded the Armor Corporation and the United Fruit Company, two food giants, to back a venture for five years at Turkey Point on the eastern seaboard of Florida and a smaller unit near Tallahassee. At the Turkey Point site, the large natural lagoons had an extra temperature advantage, because they received waste heat in cooling waters from the nearby nuclear-fueled plant belonging to the Florida Power and Electric Company, which naturally became the third partner. Large rectangular rearing ponds were laid out next to the power plant and lined with rubber, because the substrate had little clay. But clearly, further research and development still would be a key to success. For this purpose, Webber teamed up with the University of Miami, where Durbin Tabb and Alice Murphy had been actively working on marine shrimp culture for some time.

During this period in the late 1960s, Florida was obviously thought to be the place where marine shrimp farming could be developed. On the Atlantic coast at Crystal River, Ralston Purina, the animal feed company, was persuaded to use heated waters from a fossil-fueled plant to supply its pilot farm. At Panama City, with the help of Fujinaga and his colleague Mitsutake Miyamura, Marifarms Inc. established the largest site of all, enclosing two, three-hundred-acre marine lagoons and netting across a twenty-five-hundred-acre embayment. Sea Farms Incorporated of Key West set up at Tarpon Bay. The Dow Chemical and Sun Oil Companies, employing the services of Harry Cook, invested further along the coast in Texas.

However, all these large lagoon projects in Florida and the Gulf states were relatively short-lived, mainly because the lagoons were sacrosanct to sportsmen and conservationists and well protected by the law. Consequently, leasing areas for shrimp farming for many years was evidently out of the question. Yet the investors were not defeated, and for the most part, they all began to move their entire field operations to the Latin American countries, where many of them already had very large landholdings, producing everything from bananas to sugarcane. However, their migration southward signaled the end of large-scale marine shrimp farming in the United States. The Armor Corporation and United Fruit relocated their investments in both shrimp and catfish to Honduras and hired Jerry Broom from the Louisiana facility on Grand Terre Island, and two transplanted Englishmen—Eric Heald, who was doing postgraduate research at Miami, and John Spencer, the engineer from the White Fish Authority fish culture operations in the Isle of Man and Scotland. Ralston Purina picked Panama for its site, and Sea Farms set up in Honduras. Marifarms was sold to Continental Fisheries, but most of the original staff left to work in operations in Ecuador. After Harold Webber parted company from the Armor/United Fruit enterprise, he established his own company called Maricultura S.A. He found backing to set up operations in Costa Rica, with Broom once again as manager. Tabb and Heald continued research on shrimp breeding and propagation in Florida.

In addition to elevating the idea of marine shrimp farming to the boardrooms of the large corporations, the pioneer culturists in the United States made one other significant contribution to its development at that time. Fujinaga's tank technology for the *kuruma* was not applicable to many of the species of shrimp native to the Americas. Consequently, like the French researchers of AQUACOP in Tahiti, the pioneers of shrimp culture in the United States set about to adapt Fujinaga's hatchery system to their own requirements.

The propagation of juveniles of most of the indigenous American species was achieved through the 1960s by staff members of the NMFS laboratory at Galveston, Texas, most of whom were actually working on planktonic bloom and red tide studies. The early development of the "Galveston hatchery system" was started by Harry Cook, and successfully completed under Dick Neal and Wallis Clarke when Cook left for the private sector. However, their work was aided by the very practical hands of a technician called Cornelius Mock, a genuine Southerner who was wont to speak the unspeakable and who invariably got away with it. But in time, despite his affronts, the skills of Corny Mock to build and demonstrate the Galveston system were in such great demand from individuals all over the world that he aggravated the federal bureaucratic system for the next fifteen years. The personal requests for his help frequently came from royalty and ministers in developing countries through diplomatic channels, and the U.S. State Department found them impossible to refuse. Finally, the situation became so exasperating that Mock was cocooned in a new position roughly equivalent to a special roving ambassador to global shrimp culture.

With the exodus of capital investment in shrimp farming from the United States, the Central and South American countries welcomed the opportunity with open arms. Just about every country with a coastline wanted to try its luck. In addition to Mock, would-be experts from the United States, Japan, and French Tahiti flew everywhere offering advice at a price. When the first dust finally cleared, Ecuador had become the principal beneficiary of marine shrimp farming in the Americas.

10.4   Salmon farming in North America

The success with rearing Pacific salmon in captivity through the salt water stage by Tim Joyner and his group at Manchester provided impetus for two ground-breaking projects in the early 1970s. The first was the construction of a net-pen pilot farm funded by NOAA and private industry to demonstrate that both coho and chinook pan-sized salmon could be reared for the market within eighteen months. The second project was in cooperation with Washington Department of Fisheries and Wildlife to demonstrate that regional sport fisheries could be enhanced by delayed release of salmon from floating pens. Both projects proved immediately successful, technically and economically.

Watching this early work from his home on Bainbridge Island, Washington, was Jon Lindbergh. Armed with information from Joyner, he persuaded Union Carbide to form a subsidiary company, also for the purpose of producing pan-sized Pacific salmon direct for market. The company was named Global Systems. Next, with the help of Hal Goodwin in NOAA's Office of Sea Grant, the NMFS team and a marine consulting and technology company called Ocean Systems, Inc., put together the project in 1970 and anchored a pilot-scale floating net-pen system adjacent to the Manchester Station. Within two years, more than sixty-five tonnes of pan-sized farm salmon were sold to test markets. The project became the basis of the first commercial salmon farm in North America and spurred another commercial enterprise, Pacific Ocean Farms, to invest in Pacific salmon farming in Puget Sound. For several years, Lindbergh continued to manage the operations for Ocean Systems, and technical assistance was provided by Joyner and his team at Manchester. In 1979, local Seattle newspapers reported that the farm was sold by Union Carbide to the Campbell Soup Company for only $3 million, and it changed its name to Domsea (domestic seafood) Farms. This sale price was suspiciously low; the lengthy leases to operate a net-pen farm in these well-trafficked waters were probably worth that much alone, and Domsea Farms soon snapped up the lease and operations of its neighbor).

However, with growing experience of holding salmon in salt water, Joyner did not stop there. With Novotny and Mahnken, he offered the experimental facilities at Manchester to help fisheries colleagues on the East Coast for holding and reproducing Atlantic salmon. The project began with funding from the U.S. Fish and Wildlife Service (in the U.S. Department of the Interior) to propagate disease-free smolts for rebuilding the Atlantic salmon runs of many of the rivers in the New England states. However, when the smolts were ready, the U.S. Fish and Wildlife Service lost confidence about the transfer and called it off. This left Manchester with over a million Atlantic salmon smolts, and the local industry stepped in to help. By the end of the 1970s, the Manchester net-pens held the largest broodstock of Atlantic salmon in the country, and the NMFS scientists began to pass their technology to many states' fisheries authorities, as well as to the private sector.

Almost immediately, the farming of Pacific and Atlantic salmon became the subject of serious attention for research and development in both the United States and Canada. Within a decade, the two countries together were producing over twelve thousand tonnes of Atlantic salmon and three thousand tonnes of coho salmon in floating cages. But the competition did not last very long. Backed by a strong research program at federal laboratories in Nanaimo, West Vancouver, and Halifax, together with enlightened policies for farm development at both federal and provincial levels, the industry in Canada was soon way ahead. In the United States, on the other hand, the young industry was slowing to a halt as the salmon fishing industry began to implement innumerable strategies to put the competing fish farmers out of business. Unlike the fishermen of Asia and Europe, North American fishermen wanted little to do with alternative sources for their products.

(a)

(b)

Figure 10.3 United States, 1972–1974; Bainbridge Island, Washington: (a) 1972, Pacific Ocean Farms during construction off Bainbridge Island; (b) 1973, Rich Passage, Bainbridge Island with Manchester, Domsea, and Pacific Ocean Farms; (*Continued*)

(c)

Figure 10.3 *(Continued)*: (c) 1974, Domsea Farm, with complex for small cages housing pan-sized coho salmon.

10.5 Salmon in the Southern Hemisphere

The successful introduction of salmonids into ecosystems in the Southern Hemisphere was a goal of many fisheries biologists. The idea was a natural one, but attempts had been more or less fruitless for almost a hundred years. Most of the introductions used Atlantic salmon and brown trout from Europe, and rainbow trout from North America. One of the first attempts with Pacific chinook salmon from California was made in 1958 by Bill Ripley. At the time, Ripley was working for the U.S. State Department Agency for International Development, before he joined the United Nations Development Programme (UNDP) in New York. In a project initiated in southern Brazil, he planted 400 thousand eggs in the gravel of the Rio Jaquí and Rio das Pelotas, a tributary of the Rio Uruguay, and four years later some "unusual fish" were observed in the Uruguay attempting to jump the Salto falls.

According to the oceanographer Tim Joyner, this was quite logical. After they had migrated to the sea, the smolts would have been confined to the cold inshore waters of the Falkland Current, fenced in by the warm Brazil Current offshore. On their return, the great plume of fresh water from the La Plata Estuary would have led them to the Rio Uruguay and the falls at Salto. Ripley's results were exciting. In spite of his success with farmed Pacific salmon in North America, Joyner still had visions of using this ubiquitous fish to range the untapped waters of the Southern Ocean. Although all previous attempts had not met with much success, he believed that massive releases from hatcheries in southern Chile or Argentina (around the Strait of Magellan and the Island of Tierra del Fuego) would enable a major salmon fishery to be established that could be totally managed from the land.

Joyner's idea caught the imagination of John Pino at the Rockefeller Foundation, who was eager that the foundation's new offspring, the International Centre for Living Aquatic Resources Management, would quickly make an impression on international fisheries management. Pino also believed that should the project be successful in creating a new fishery where there was none, it would be an opportunity for model management of an international fishery, starting from absolute zero. Consequently, the Foundation funded Joyner and a team of aquaculture and salmon experts to visit Chile in 1975, select sites for hatcheries in the southernmost Magallanes region, and put together a program for major international funding by the Inter-American Development Bank.

Because of the political and economic isolation of Chile in the years following Salvador Allende's presidency, which ended in 1973, the program they formulated for "Seeding the Southern Ocean with Salmon" was never funded. But several missions by Joyner and his team of salmon specialists between 1975 and 1978 accomplished three things that would have a lasting impression in Chile. First, they persuaded an existing cooperative program of the Japan International Cooperation Agency and the government to switch from Japanese strains of cherry, chum, and pink salmon to American coho and chinook salmon, which would be more conducive to the coastal waters of Chile. The program had been working since 1969 on the introduction of salmon from a hatchery in Coyhaique. Second, they persuaded Aliaky Nagasawa and Pablo Aguilerra, the joint managers of the hatchery at Coyhaique, not to release the presmolts from the hatchery immediately into the Rio Simpson, but to hold them through smoltification, using floating cages located near Aysen Fjord. Third, they persuaded Fundaçion Chile to support research and development of the potentially important field of aquaculture. Fundaçion Chile had just been created in 1975 to handle several millions of American dollars belonging to the International Telegraph and Telephone Company, which had been confiscated by President Allende and subsequently returned after his government was overthrown. The newly appointed director of Fundaçion Chile, Robert Cotton, was looking for new technology projects to support and welcomed the idea of aquaculture—particularly as a potential industry for the remote southern part of the country.

The rest, as they say, is history. Lindbergh arrived in Chile almost immediately after Joyner and his team. Working again through Union Carbide, which had mineral rights and capital in Chile, Lindbergh led the way in the introduction of salmon farm technology in marine waters using the Ocean Systems model. Other enterprises quickly followed, all of them starting off by producing coho salmon, but later substituting Atlantic salmon and rainbow trout as more eggs became available on world markets. For several years, Fundaçion Chile supported

(a)   (b)

Figure 10.4 Chile, 1977: (a) Coyhaique salmon hatchery, Chile; (b) first release of young salmon into the Rio Simpson. (Courtesy Aliaky Nagasawa.)

research and development, not only with salmon, but also with several valuable species of shellfish. By the end of the 1980s, starting from nowhere, Chile became a major producer of farmed salmon, with over 15 thousand tonnes of Atlantics, 18 thousand tonnes of Pacifics, and 8 thousand tonnes of rainbow trout. A decade later, it was producing over 227 thousand tonnes. Its exports were earning over $800 million and displacing home-grown salmon on the American markets.

10.6 Atlantic salmon farming in Europe

Throughout the early years, the pioneers of the embryonic industry of salmon farming in Norway and in neighboring European countries, such as Scotland and Ireland, had little outside help. It was not until their products began to trickle onto the markets of Europe early in the 1970s that everything rapidly changed. Their respective governments woke up and began to back the promising technology.

Norway led the way in providing supportive legislation and offering investment grants through its Regional Development Fund. The government also financed key research and development, and making use of the public university system, set about designating research centers to focus attention on the disciplinary needs of the burgeoning industry, and to fill them with the top scientists by providing them joint positions. But delegating responsibility for research on each new and important field was not easy. There was significant competitive

Figure 10.5 Chile, 2003; aquaculture of Atlantic salmon: (a) a farmed Atlantic salmon; (b) harvesting; (c) processing. (Courtesy of SalmonChile.)

rivalry between the two concerned directorates, Fisheries and Agriculture, and between their principal researchers. The delicate political situation was resolved by the government's agreeing to allow some entirely new facilities to be built for the job. In 1971, Fisheries funded its Institute of Marine Research to build the Matre Aquaculture Research Station near Bergen, which was to be directed by Dag Møller. Then, in 1974, because of the persuasive fund-raising talents of Harald Skjaervold, the Agriculture Directorate built the Institute of Aquaculture Research (Akvaforsk) at Sundalsøra and refurbished the Institute of Animal Genetics of the Norwegian College of Agriculture at As, which was directed by Trygve Gjedrem.

The new breed of fish farmers in Norway was also quick to organize. First they formed the Norwegian Fish Farmers Association in 1970, and this was followed by the Norwegian Fish Farmers Sales Organization in 1978. The farmers gave the right of all first-hand sales of farmed fish and shellfish to the Sales Organization, thus maintaining stability in the industry at home, while satisfying the growing markets in Europe for their popular high-quality products. Although the Sales Organization would subsequently come to grief fifteen years later, the organization of the industry made commercial salmon farming in Norway a spectacular success. Records were broken year after year, and by the end of the century, Atlantic salmon production by the industry was valued at well over $1 billion.

10.7 Private ocean ranching in North America

Public financing has monopolized anadromous ocean ranching operations all over the world, primarily for the five principal Pacific salmon species. The greatest concentration of public effort was in the United States, particularly in the Pacific Northwest, but the first hatchery in Oregon was in fact private. It was built in 1877 on the lower Rogue River by a salmon packer called R.D. Hume, who operated it privately for eleven years until it was financed by the state.

About a century later, the state of Oregon passed legislation to allow limited private ocean ranching. It was the first state to do so, and many millions of dollars were invested in several private farms along the central Oregon coast. One of the first into the fray was Oregon Aqua Foods, already a pioneer in pan-sized salmon production, and it recruited Bill McNeil from Alaska to manage its development in ranching. Subsequently, the company was taken over by the Weyerhauser Company, the West Coast timber giant. Production through private ocean ranching was mostly focused on coho and chinook salmon, although some unsuccessful attempts were made to establish chum salmon runs from nonindigenous stocks. Private hatcheries released millions of smolts, and the average survival was exceptionally high. However, many of the returning fish were exploited by commercial trollers and recreational fishermen, contributing in some years over 30% to their harvests. Such losses made these operations unsustainable, and all the private farms had closed down by early 1980s, leaving ocean ranching once more to the state's thirty-four public hatcheries.

Alaska, in contrast, began to operate two types of ocean ranching in 1970s. These were the publicly funded enhancement programs, which included some federal funds, operated primarily by the Alaska's Department of Fish and Game through the Fisheries Rehabilitation, Enhancement, and Development Division, and privately funded enhancement programs of the Private Non-Profit Aquaculture Associations (PNPs) legalized in 1974 to operate hatcheries in the state. The PNP system became more operable when 1976 legislation created regional aquaculture associations, which were essentially cooperatives made up of commercial fishermen and other user groups, such as processors, sport fishermen, subsistence fishermen, and local or regional municipalities.

The public program began in 1973, and the fisheries department built and operated thirteen hatcheries. The PNPs constructed its first hatchery in 1976, and eventually a total of fourteen hatcheries scattered throughout southeast and western Alaska. The PNP group included eight regional aquaculture associations and twelve nonregional corporations, all with the common objective of providing harvestable salmon to specific fishing communities. Most divided the yield as follows: 70% of fish for the commons and 30% for cost recovery and broodstock. By 1985, over one billion eggs were collected from all state and nonprofit stations, and the estimated return was nineteen million adults to Alaskan waters.

In addition to the five species of Pacific salmon, state hatcheries released rainbow trout, grayling, steelhead trout, sheat-fish, and Arctic char.

The program of the PNPs increased steadily in size and regularly made annual releases of over one billion juvenile salmon. This was about 76% of their permitted capacity, and adult returns of Pacific salmon climbed as high as 42 million. Some of the PNP hatcheries established record-breaking numbers. The Prince William Sound hatchery at Port San Juan, for example, was able to take over 128 million eggs from pink salmon returns in less than a month, and a Solomon Gulch hatchery near Valdez, in Prince William Sound, had a record take in 1993 of 231 million eggs of pink and chum salmon. The 1993 release of pink salmon from this hatchery was 142 million juveniles, with a projected return in 1995 of some 5.5 million adults (4%). For the region, this number would have represented a major increase in production. Returns from wild runs of pink salmon in Prince William Sound are generally between 0.1% and 0.5%.

This large increase in production, particularly of the small and less expensive pink salmon, created a logistical problem for the processors and a financial problem for the marketers, because the species is primarily used as a canned product. Moreover, the price paid to the fishermen declined over 66% to less than 30 cents per kilogram. Thus, in spite of the abundance of fish, some commercial fishermen could not make a profit harvesting pink salmon.

10.8   Salmon ranching in Japan

Salmon propagation for release into the open ocean was started in Japan in 1888, with the construction of the Chitose Central Salmon Hatchery Station in Hokkaido on the Ishikari River. The hatchery was modeled after the Bucksport Maine Station, on the northeast coast of the United States. It was also fitted with a trap to catch the returning fish, which was a copy of a fish wheel operated by native Americans on the Columbia River. Just after the turn of the century, there were more than fifty salmon hatcheries in Hokkaido, most of which were privately operated for the sale of spent fish, rather than for the sale of the progeny back to the government. But with the general disinterest in the state of rivers and the growing industrial pollution, many of these small family-owned hatcheries were inefficient and run down. In 1934, the majority were taken over by the Hokkaido Government.

Even under the control of the government, there was little or no improvement in the Hokkaido salmon fisheries for another fifteen years. In 1951, Richard Van Cleeve, the dean of the College of Fisheries at the University of Washington, was called on by the American Occupational Forces to carry out a complete survey of the salmon rivers in the country and the national salmon industry as it existed at the time. His report made sorry reading. Although it was well known that the short length of most Japanese rivers and their flow rates were not entirely conducive to salmon breeding and propagation, it was obvious to Van Cleeve that most of the natural habitat was almost beyond recovery. The electric power companies and the large industries wielded total control over the use of water in the rivers and ignored any laws that attempted to protect the environment and the salmon. Even though some river barriers had fishways, the water flow was insufficient to provide any passage for the fish. Although Van Cleeve proposed a program for widespread rejuvenation of the natural habitat and the construction of fish passageways, it was ignored.

The Japanese markets are the largest consumers of salmon in the world, and the country now manages and controls its salmon fishery with great care. However, in 1951, the natural fishery for salmon in Japan was essentially perishing. The authorities had to do something, and quickly. In 1952, the entire salmon propagation system was incorporated into a national salmon development program based on artificial propagation of all salmon in hatcheries. A few hatcheries remained in private hands, but these were actively involved in the program, and their operations were closely monitored by the organization created for salmon propagation. This was the National Salmon Hatchery Service, which was to have tremendous long-term benefits for the offshore and coastal salmon fishing industries.

Based on the annual propagation and release of millions of chum and pink salmon each year, the annual catch by the Japanese high-seas fleet rose steadily through the 1960s to over 120 thousand tonnes. But with the adoption of the Extended Economic Zone to two hundred miles by Japan and most other nations in 1973, the high-seas fishery began to drop alarmingly.

The salmon fishery in Japan also supported large numbers of coastal communities, which since 1948 had been given the rights to catch the returning fish in set nets, provided they formed fishing cooperatives. These cooperatives had become large and powerful, and when united with the large numbers of high-seas fishermen, they formed a significant public voice, which they used to demand a solution for their declining fisheries. Salmon ranching was the answer. Working together with the two groups the National Salmon Hatchery Service stepped up production releases year after year. Over two billion juvenile salmon, of which about half are chum, are released annually, with records of almost twenty million chum salmon harvested in the coastal set net and river-mouth fisheries of Japan.

Commercial salmon farming, on the other hand, is quite recent. Like many others, Akimatsu Koganezawa and his colleagues began in the 1960s with rainbow trout in salt water in Hokkaido. Small farms grew up, and by 1975, production was over three hundred tonnes. Then, faced with the terms of the new Extended Economic Zone and fishing quotas, the Nichiro Fishing Company began to invest in commercial production of Pacific salmon species, eventually selecting coho. Production increased rapidly, and by the end of the 1980s, the country was harvesting over sixteen thousand tonnes of farmed fish. But it was not to increase. With the growing production of Atlantic salmon, and cheaper production of coho salmon in Chile, national farmed production had reached its peak.

10.9   The big interests of "big business"

The rapid expansion of aquaculture in the 1970s did not go unaided by the private sector, particularly by "big business"—those at the level of Fortune 500. Indeed, many of the large international corporations that had already been investing in research in the 1960s were by then eager to extend into farm production. Others that kept an ear to the ground for new investment opportunities were readily persuaded that the new field was ripe for the taking. As usual, large corporations never did anything by halves, and their need for a healthy bottom line for shareholders led some of them into vast schemes with substantial capital and operational costs. If such projects did not materialize in three or four years, then they were terminated or sold. Some of the corporations involved had little or no practical experience in agriculture of any kind, did not understand the time-scales involved for development and were not institutionally accustomed to the levels of risk involved in aquaculture. Thus, the eventual outcomes could have been largely predictable.

Many of the early investors were American businesses. One of the first was a joint effort of two giant corporations, Armor and the United Brands. Using technology for the culture of marine shrimps indigenous to the Gulf of Mexico, developed by scientists at the Institute for Marine Sciences at the University of Miami, Florida, and the federal laboratory of the old Bureau of Commercial Fisheries at Galveston, Texas, the team began operations in Florida in lagoons fed by cooling waters from a nuclear generating station. However, when some agents in the water proved to be toxic, the project was moved south. Because United Brands had large landholdings in Central America for bananas, the corporation decided to convert some of its coastal plantations in Honduras to marine shrimp production. Other large corporations interested in backing marine shrimp development at the time were the Dow Chemical Corporation (chemicals), the Sun Oil Company (petrochemicals), and Ralston Purina Company (animal feeds).

A bigger draw at the time, however, was farming of freshwater prawns, which had been achieved successfully by Takuiji Fujimura and his state team in Hawaii. Again, Dow Chemical Corporation and Ralston Purina Company were attracted, but there was also interest from the Weyerhaeuser Corporation (forest products), General Mills Inc. (foods), C. Brewer and Company (sugar and chemicals), AmFac Corporation (sugar and retail stores), the Coca Cola Company (soft drinks), Walt Disney Productions (entertainment), and the Syntex Corporation (pharmaceuticals). Others, such as Consolidated Mills (agriculture and animal feeds), were more interested in catfish. Companies with certain exploitation rights of natural resources, both in the United States and overseas, turned to Pacific salmon and trout. The Weyerhaeuser Corporation also purchased Oregon Aqua Foods from Lauren Donaldson (the doyen of pioneer trout culturists at the University of Washington) and his son to carry out both ocean ranching and pen-rearing of Pacific salmon. CrownZellerbach (timber and paper) and Union Carbide Corporation (minerals) also invested in salmon farms, whereas the Inmont Corporation purchased the Thousand Springs Trout Farms in Idaho and Long Island Oyster Farms in New York.

Many large private power generators and oil companies in the United States made large investments in the new business of aquaculture as well, or became noninvesting partners in enterprises that made suitable use of their various resources. Several of these companies had mixed motives. Although sharing in the profits from the production of marketable seafood would be a bonus, the main motive was to bask in the good publicity of being a caring company, active in a benign pastoral activity such as fish and shellfish farming. These efforts to demonstrate the beneficial side of their industries were mostly to appease the growing environmental movements in the country concerned with thermal enrichment, waste disposal, and general pollution. Pacific Gas and Electric Company offered its warm waters at Moss Landing to International Shellfish for the culture of oysters, clams, and abalones; Florida Power and Light Company and the Houston Power and Lighting Company supported marine shrimp projects in Florida and Texas; Cities Services Oil Company went into catfish culture; and Humble Oil Company turned to culturing several marine fish native to the waters along the Gulf Coast.

A similar trend could be seen among the large, multinational corporations based in Europe, but on a smaller scale, because there was no temptation in European climates to follow the lure of large, quick profits from marine shrimp and freshwater prawns. Norsk Hydro, the Norwegian industrial conglomerate, purchased 50% of the shares of Mowi A/S of Bergen and went into the industrial production of Atlantic salmon in Norway. Salmon was also the interest of the Unilever Group, and through its new subsidiary called Marine Harvest Ltd., it built research and development centers in Findon near Aberdeen for crustaceans and flatfish, and at Loch Ailort on the west coast of Scotland for rainbow trout and Atlantic salmon. The British Oxygen Group purchased Shearwater Fish Farming Ltd. in north England and invested heavily in research and development to study the culture of rainbow trout and marine shrimp at very high oxygen levels. British Petroleum purchased 49% of Fish Farm Development International, another salmon rearing and consulting enterprise started by Ian MacFarlane.

Other, perhaps unwitting investors in supporting the developing technologies were the public utilities. In 1966, the South of Scotland Electricity Board first agreed to allow the White Fish Authority to carry out its research program to develop flatfish farming in the warm waters discharged from the Hunterston Nuclear Generating Station in Ayrshire. By the mid-1970s, the site was leased over to Golden Sea Produce Ltd., operated by Guy Mace and his father for the production first of eels, and then of flatfish. Down south, the Central Electricity Generating Board in England agreed to provide water from Hinkley Point nuclear plant to Marine Farm Ltd., another oyster and eel rearing farm pioneered by Maurice Ingram.

Many businesses, both large and small, nurtured the burgeoning aquaculture field in the 1970s. Few, however, would still be playing the game 10 years later.

Chapter 11

The Rise of the Institutions (1970–1980)

Abstract

In the 1960s, aquaculture was polarized. Aquatic animals and plants were farmed using traditional practices in Asia and Africa. Although Western countries had few species, they had blossoming aquaculture technology. No infrastructure existed to bridge between developed and developing worlds. The U.S. Peace Corps and British Voluntary Service Overseas provided field technical ambassadors in the 1960s and 1970s. Western bilateral assistance organizations supported aquaculture research, education, and vocational training. The Oceanic Institute and Rockefeller Foundation formed the International Centre for Living Aquatic Resources Management, realizing aquaculture production, food, and development in Oceania. The United Nations Food and Agriculture Organization's Global Program held meetings such as the 1976 "Technical Conference on Aquaculture" in Kyoto. Private and national consultancy groups expanded global aquaculture. Communications matured with the World Mariculture Society (1969), and new journals and books increased awareness among universities, institutions, and investors. New aquaculture insurance secured stability and growth while necessitating improved farm standards.

11.1 Introduction

Toward the end of the 1960s, aquaculture was distinctly polarized across the geographic map of the world. On the one side, primarily in the developing countries of Asia and Africa, a wide array of aquatic animals and plants was being produced and sold in the marketplace. There were countless thousands

The History of Aquaculture. By C. E. Nash. Published 2011 by Blackwell Publishing Ltd.

of skilled practitioners dependent on farming for their livelihoods, but mostly they used the same simple practices that had been traditional for centuries. On the other side, predominantly in the developed counties of the Western world, only a handful of species were being raised, and farmed products in the market were something of a novelty. Few people had any real experience in farming fish or shellfish for a living. But with the excitement of space-age achievements, there was renewed political interest in exploitation of the resources of the sea, and consequently, most developed countries were beginning to invest heavily in aquaculture research and development. Results were solid if not spectacular, and soon the West was sitting on several new production technologies and technical improvements. However, the leading countries did not have the critical mass of producers to use this knowledge to any real advantage or consumer markets demanding of the products. These lay overseas.

The need was for technology transfer, but there was no existing infrastructure bridging the gap between the developed and developing worlds. Furthermore, the gap was distinctly wide, and the bridge required three spans. One was for educators with qualifications and hands-on experience in aquaculture who could teach trainers; another was for experienced trainers who could instruct extension workers; and the third was for extension workers with applied skills, who could carry the technologies to the farmers in the field. Fortunately, many countries were not prepared to wait for the orderly building of institutional infrastructure to meet the need, and for better or worse, it was the extension workers who first volunteered to leap into the fray.

11.2   The voluntary services

Through the late 1960s and early 1970s, the voluntary services provided the front line of technical ambassadors who spent two or three years living in the field with fish farmers and their families, and extended information about new aquaculture technologies. The majority of the old colonial countries of Europe had long-established schemes for sending trained graduates overseas, but the programs were not entirely voluntary. They were usually operated through a department in the ministry of foreign affairs and linked to specific national educational institutions. Overseas service was therefore recognized as a part of a student's higher education; in France it could replace military conscription. Some of these institutions expanded to cater to specific technologies. In the Netherlands, for example, the Agronomic University of Wageningen was a principal link in the transfer of agronomic technologies to the old Dutch colonies and had been outposting students for decades. In Belgium, the link was with the Research Station of Waters and Forests at Groenendaal, with facilities for fish-culture students and with medical schools for students to work on water-related tropical disease projects.

An early contributor to the global development of modern aquaculture was the Voluntary Service Overseas (VSO). The VSO had been founded in the United Kingdom in 1958 as a charity with a mission to send aid mainly to former colonies in the form of expert volunteers. In time, the VSO would recruit volunteers from all member countries of the European Community and from Canada and send them almost anywhere in the world. However, by far the most active and influential volunteer service for aquaculture extension was the U.S. Peace Corps. President Kennedy first broadcast his rallying call in 1963 for young men and women to promote world peace and friendship through community involvement and personal sacrifice. The principal themes of his new Peace Corps were education, health, and agriculture—all of which were good excuses for young and enthusiastic volunteers who were eager to work overseas to try their hand at fish farming.

Aquaculture was clearly a field that fit right into the purview of bilateral assistance, but it would be the mid-1970s before the voluntary services became an effective force. With the exception of the Dutch and Belgians, the early volunteers were mostly college graduates with no experience or artisanal skills in aquaculture. Armed with little more than their degrees and diplomas in biology or agriculture, and a sense of adventure, the young men and women had to be exceedingly resourceful. For the early recruits into aquaculture, it was "on-the-job" training. They had phrase books to teach them the local languages and guide books to advise them on social customs, but they had no technical manuals to help them improve the lot of the fish farmers.

Not surprisingly, the first volunteers to return were coerced into becoming the trainers for the next year's class of green recruits. In addition to holding simple initiation and training courses, the earliest pioneers were persuaded to write basic training manuals and to produce any other training aids they would have found useful. They put together low-technology field kits to enable their successors to make simple tests for water quality parameters, and in time, they would assemble a basic aquatic animal health kit.

By the end of the 1970s, it was apparent that the work of the aquaculture volunteers was greatly appreciated by the primary producers on the farms. In 1975, the Peace Corps had requests for 40 volunteers to assist fish farming projects, mostly in Central America and Africa. By 1977, the number was 101, and the following year, it was over 250. By that time, the requests came in from all over the world. Fortunately, to meet this demand, other organizations had started to help the Peace Corps by giving the recruits some basic training. The University of Oklahoma and Auburn University provided premission technical courses in freshwater fish farming, and the Oceanic Institute in Hawaii provided similar courses in marine fish breeding and hatchery technology.

Without doubt, the volunteer institutions from Europe and North America were responsible for accelerating the global development of aquaculture between 1965 and 1980. They recruited the pioneers who could carry the technological information about modern aquaculture across the continents of Asia, Africa, and Latin America and apply it right at grassroots level. Yet, they were also organizations that were largely ignored by the national governments, because the local functionaries were more concerned with the projects of the multilateral and bilateral assistance agencies that provided cars and study tours. Consequently, the voluntary services had to find tough young men and women who had to live the life of subsistence farmers and their families. There were no expense accounts for the comfort of hotels.

Truly worthy of inclusion in the list of volunteer institutions that helped pioneer aquaculture development at grassroots level are many relief organizations. These included, for example, the Catholic Relief Services, as it expanded into Africa, Asia, and Latin America in the 1960s and 1970s and added agricultural initiatives to its roster of existing projects, such as CARE (Cooperative for Assistance and Relief Everywhere, Inc.). However, among the most influential of the volunteers who were willing to add to the responsibilities of their own lives by introducing fish farming around the world was Father Jan Heine. Heine was a Jesuit priest from the Netherlands who created the Tilapia International Foundation. The rather grand name of his foundation belies its role as a network of rank and file volunteers introducing St. Peter's fish to the rural poor of the Third World countries. In his unassuming way, and relying totally on small donations garnered by the Tilapia Foundation's administrators in Europe, Heine and his extensive network of volunteers probably contributed more to the practical production of fish to be eaten by real village communities than all the million-dollar rural development schemes for fish farming created by the international assistance organizations put together.

The voluntary service institutions would continue to play an important part in the extension of modern aquaculture for another twenty years, but the policy of the individual organizations slowly began to change. Sensibly, the majority opened their doors to adult volunteers, thus drawing on men and women who were already well experienced in all manner of professional and artisanal skills. This was not possible in the late 1960s and early 1970s, because there were very few individuals available who had any useful aquaculture experience other than training in fisheries biology. Fortunately, the lack of experience has never been a detriment to youthful ego, and there was no shortage of volunteers, many of whom would continue to play greater roles in the cause of aquaculture development on their return home.

11.3   The international donor community

Immediately in the wake of the voluntary services that filled the gap for extension of aquaculture at local levels in many parts of the world, several bilateral assistance organizations of the West stepped up to support developing countries in aquaculture research, tertiary education in a range of related aquaculture fields, and vocational training. Like the voluntary groups, their well-meaning programs took time to become operational, because apart from the Dutch and Belgian organizations, most were constrained by the lack of national experts and institutions capable of supporting specific aquaculture activities, especially in tropical and subtropical environments.

The first, and probably the most effective, new organization that began to work in aquaculture was the International Development Research Centre. It was created by the government of Canada in 1970 as a public corporation. Its focus was exclusively on research and research information designed to adapt science and technology to the needs of developing countries, and the needs were specifically confined to a few key fields—of which fisheries was one. But its geographic purview was essentially worldwide, and therefore, within a short period of ten years, it made some outstanding contributions to aquaculture development.

The Research Centre operated through a number of regional offices, of which the largest and most relevant for aquaculture was one in Singapore. Within each well-defined field of interest to the Research Centre, the individual program directors had a reasonable amount of latitude. It was fortunate for aquaculture that two of the directors at that time were Bert Allsopp and his successor Brian Davy. Both were sympathetic toward aquaculture, and each recognized that only persistence would be rewarded. Between them, they organized a series of activities that enhanced a number of research fields throughout the region for almost two decades. In addition to funding specialized workshops for researchers on fish culture, reproductive physiology and induced breeding, fish quarantine and diseases, and economics, the two contracted for the production of a range of pocket-sized publications that the organization distributed free of charge, both far and wide. They also made a film on milkfish farming. But the Research Centre's greatest contribution was the funding of innumerable promising individuals who were nominated by Allsopp and then Davy for overseas fellowships for research studies in aquaculture-related fields. As time went on, the connotation of "overseas fellowships" became "regional fellowships," as Africa, Asia, and Latin America established their own institutions academically qualified to give postgraduate degrees and capable of providing advanced facilities for aquaculture research.

Another early effective force in aquaculture research throughout the 1970s was the International Foundation for Science. This Swedish organization was established in 1972 with a commitment to support research in developing countries, particularly in Asia. It provided modest grants of about $5,000 to provide young research scientists working in their own countries with some necessary equipment and materials. But that was all. The International Foundation did not pay a stipend or expenses. However, that which each grant may have lacked in flexibility for any of the recipients was more than offset by the number of recipients awarded grants by the foundation each year. At any one time, up to fifty individuals in the region could be carrying out their research supported by the foundation. Furthermore, the organization advanced regional cooperation and encouraged technology transfer by hosting sponsored regional meetings of all grant recipients, and then publishing the proceedings in international peer-reviewed journals. The first such meeting was held at the Universiti Sains Malaysia, Penang, at the end of 1978, and the proceedings were published in 1980 by Elsevier Science B.V. in *Aquaculture*. By 1987, the International Foundation for Science had made its thousandth research award in the field.

Other bilateral assistance organizations were soon ready to follow suit in supporting higher education and technical training in aquaculture. Each of them had been active in traditional capture fisheries projects for a decade or more;

however, supporting projects in the new field of aquaculture was not an easy objective to achieve, due to the shortage of national experts to carry the projects through. So, for most of the organizations, the elements of research, education, and training were integrated into specific projects, and responsibility for their conduct was relegated to institutions in the home country. The first to move were the old colonial countries of Europe. The Dutch had the immediate help of the Agriculture University at Wageningen, and the Belgians called on the Research Institute of Waters and Forests in Brussels and the State University of Ghent. The French used their traditional resource, the Technical Centre for Tropical Forests, for freshwater fisheries projects and relied on its oceanological centers in Brest and Tahiti for work in the marine field. The British had to rely at first on some old colonial hands attached to the Fish Section of the British Natural History Museum before it began to call on government fisheries laboratories and the quasi-government White Fish Authority. In the United States, most aquaculture projects of the U.S. Agency for International Development were linked favorably to Auburn University, and capture fisheries projects were connected to the University of Rhode Island.

Some countries new to the field of bilateral assistance had national institutions that were just as well qualified in their familiarity and experience with aquaculture technologies, but had something of a language barrier for potential postgraduates or research fellows working in the home country. Both Norway and Japan, for example, quickly realized the potential language difficulties for non-national students. Consequently, instead of awarding fellowships, they concentrated on very practical short-term training at their principal centers of research, in government production facilities, and on private farms. The results were equally good, if not better than those of the British, French, and Dutch. The Norwegian Organization for Research and Development would become one of the most circumspect proponents of aquaculture, placing its bilateral assistance in well-chosen and altruistic projects that had broad impact for everyone. Equally as effective was the Japanese International Cooperation Agency through large training schemes in Japan for both South Americans and Asians in parallel with bilateral capital assistance projects.

By the end of the 1970s, most of the donor countries had strengthened their resource base of experts and institutions. National experience for tropical and subtropical aquaculture had been built up by on-the-job training as well as by voluntary service in regions such as Asia, Africa, and Latin America. The experience base was so enriched that the following decade would see a high level of competition between national institutions and among groups of consultants for the many contracts being made available for aquaculture by the donor community, and for the honeypots of the international, multilateral assistance organizations waiting just around the corner.

11.4   Southeast Asia Aquaculture Development Center

By the mid-1960s, Japan was recognized as the world leader in new aquaculture technologies and, having established a strong national industry, was ready to pursue programs of international cooperation in the region. At the end of the 1960s, through its newly formed Japanese International Cooperation Agency, Japan financed a team of national aquaculture experts under Director of Fisheries Katsuo Kuronuma at the University of Tokyo to identify a major site in the Philippines for research and development, and for training in marine shrimp culture for the benefit of the five Southeast Asian nations under the newly formed intergovernmental organization, SEAFDEC (see also in Section 10.2).

The member nations that formed SEAFDEC in 1968 were Indonesia, Malaysia, Philippines, Thailand, and Singapore. Although the Japanese Agency contributed the lion's share of its early funding, the U.S. Agency for Regional Economic Development, headquartered in Singapore, also contributed financial assistance in the early years. The SEAFDEC organization was structured into three principal departments: one for training in fishing technology, which was to be located at Patnam in Thailand; another for marine fisheries research and development to be based at Changi in Singapore; and the third for aquaculture development to be initiated in the Philippines.

Kuronuma's aquaculture study team settled on a site in Leganes, near Iloilo, on the island of Panay. Coincidentally, the site was directly adjacent to another pond complex being developed by Homer Swingle and his team from Auburn University, with financial assistance from the U.S. Agency for International Development for research and development in the culture of milkfish. At Leganes, the Japanese designed a small laboratory and thirty acres of shrimp ponds. The ponds were dug more or less manually in less than a hundred days through the wiles of a Buddhist philosopher called Queron Miravite, and the Aquaculture Department was on its way. During the next decade, with his acumen for publicity—and a direct telephone line to President Marcos—Miravite would obtain national funds to expand the facilities at Leganes, build a vast new research and training center down the road at Iloilo City, and establish a number of field sites throughout the Philippines.

The growth and expansion of the SEAFDEC Aquaculture Department in the Philippines under Miravite began to dominate the entire regional program of the organization. Soon the Aquaculture Department developed into a key regional institution in its own right. It rapidly became a magnet for broader cooperation with other agencies and was a direct recipient of international funds, thus almost bypassing the central organization—much to the chagrin of the other four member nations, each of which was supposed to have, eventually, a fisheries center of some type. A fourth department in SEAFDEC, which was for fisheries resources and management, would not be created until the early 1990s and headquartered in Malaysia.

Notwithstanding the many international and national jealousies within the SEAFDEC organization about the success of the Aquaculture Department, its presence in the Philippines was responsible for an extraordinary amount of regional education and training in aquaculture skills. Through a favorable affiliation with the University of the Philippines, the department was able to offer a postgraduate degree course in aquaculture. This had the added advantage of being able to offer places to students from outside the SEAFDEC's five member countries. Consequently, the degree course was extremely popular, not only throughout the Southeast Asian region but also for students in Latin America and Western Asia. For the next quarter century, the Aquaculture Department of SEAFDEC in the Philippines was a significant factor underpinning research and higher education in aquaculture in many developing countries.

11.5   The World Mariculture Society

In the shadow of the 1960s, a group of wildlife scientists and culturists in the United States had been formed into a subcommittee for mariculture as part of the work of the Gulf States Marine Fisheries Commission. The first informal meeting was held at the Marine Research Laboratory on Grand Terre Island in Louisiana in January 1969, and it was clear that the scientists had many technical interests in common and much to talk about. This was quickly followed in September of the same year by a more organized session at the Gulf Coast Research Laboratory in Ocean Spring, Mississippi. A steering committee, which had been established to structure the group but "at a strictly shirt-sleeve level," proposed that a new proper society should be formed, predominantly for anyone from the southern states interested and working in mariculture. A charter and by-laws were prepared, and the group needed a name. According to those present, Gordon Gunter, the director of the laboratory, stated categorically that there were many international societies in existence, and with such a precedent, this one should be called the World Mariculture Society. And that was that.

The name of the society was logical to all those present at the time. All were interested in marine fish and shellfish culture, and to those who live in the southern United States, that region is a world of its own. However, in time, as the World Mariculture Society grew in strength and became truly international, the name became a distinct problem. Although membership was open to the world, it was always painfully clear that the society was controlled and dominated by the Americans. Further, mariculture was but one small part of modern aquaculture, and the word "mariculture" was not popular among the increasing number of fish culturists. Although the etymology was undoubtedly correct for marine water culture, it was not a useful word in that its definition by derivation did not assist the definition of other parallel activities, such as freshwater culture and brackish-water culture.

Etymology was obviously not a strongly inherited suit for those modern fish culturists. The pioneers of the nineteenth century, who were well trained in Latin, Greek, and the scientific orderliness of Linnaeus, adopted "pisciculture" for the work with fish, and "ostriculture" for work with oysters. For the generalists who followed, not trained in the classics, "fish farming" and "fish culture" were more commonly used, but not entirely accepted by all because of the increasing interest in crustaceans and mollusks. "Finfish culture" and "shellfish culture" were therefore tried for a period, primarily in North America, but without much conviction. The underlying need was for one word that would include all cultures of aquatic animals and of aquatic plants, as well. The most obvious word was "aquaculture," the cultivation of the waters in parallel to "agriculture," the cultivation of the lands. However, even this was not easily adopted, and for a time, the alternative spelling of "aquiculture" was widely used for no better reason than it just looked closer to the word "agriculture."

Notwithstanding any future dispute over its name, the new World Society opted for the word "mariculture," reasoning that its members were practicing the cultivation of the sea. The name was used in the articles of incorporation that were registered in Florida at once. The First Annual Workshop was held in Baton Rouge in Louisiana in February 1970 and was attended by about two hundred scientists and culturists from the region.

The ensuing story of the society through the 1970s is one that mirrors the rapid expansion and maturity of aquaculture worldwide. By the end of the decade, membership had increased to over one thousand and was thought to be widely spread among some twenty countries. In addition, in 1976, many of the European members had reformed under the banner of the European Mariculture Society, quickly changing the name to the European Aquaculture Society before becoming the first affiliate of the parent organization. By the end of the millennium, the membership of the global society numbered over four thousand individuals dispersed in some ninety-four countries.

The organization did change its name to the World Aquaculture Society in 1986. The undercurrent from members in the society to change the name had been moving for some time and for several reasons. A number of members could not come to terms with either the words "world" or "mariculture" in the name of the organization, because neither was applicable. The society was neither worldly nor did it restrict membership solely to those working in the field of mariculture. The members preferred the more common terms "international" and "aquaculture." But the more compelling reason was that the society was the dominant association of professionals in the United States, and as an international organization by constitution, it was not an effective *national* body.

In the United States, where political lobbying is almost a national pastime taught as part of the curriculum in the public school system, any association of professionals is useful to pressure politicians for financial support for research and development or for any other self-interest. The real producers, under the banners of the Catfish Farmers of America and the U.S. Trout Farmers Association, were actively lobbying for aquaculture from the start, but they were not supported in science and technology by any professional group—other than a token voice from the American Fisheries Society that had its own agenda to pursue. Thus, the truly American interests in aquaculture needed to be separated out of the parent World Mariculture Society body, so that together with the producers, there could be a well-rounded national force representing the growing aquaculture industry in the United States.

The structure of the society was threatened for a time, because some members wanted to break away and form a national society, but cooler heads prevailed. After details of affiliation were finalized, the articles of association and by-laws were changed. The new name was the World Aquaculture Society, and the way opened for the formation of an affiliated United States chapter (United States

Aquaculture Society) without threatening the original membership. Other affiliates soon followed, such as the European Aquaculture Society, the Aquaculture Association of Canada, and the Caribbean International Aquaculture Association, and there were expressions of interest in the Southern Hemisphere from Latin America and the Australian Mariculture Society. Clearly, news travels slowly "down under" – that is, the Australians were not aware of the political challenges of choosing that name.

The rapid growth of the society had much to do with its timing. Throughout the 1960s and 1970s, more professionals were entering the new and interesting field, but there was no scientific and technical network through which to communicate. The World Mariculture Society, probably because of its name more than anything else, offered such a forum and attracted many members worldwide, most of whom had little opportunity to attend any of its meetings. Unquestionably, if Gunter had said, "Let's call ourselves the Gulf Coast Mariculture Association," history would have been very different.

11.6   The International Centre for Living Aquatic Resources Management

In 1968, the private Oceanic Institute in Hawaii had received a major research grant from the Rockefeller Foundation for the development of brackish-water fish cultivation. The Rockefeller Foundation's interest was in realizing the potential of aquaculture for the production of food for the poor of developing countries, and the common gray mullet was believed to be capable of filling this role if it could be bred and propagated in captivity. Although the program was technically successful, the results each year were inconsistent, and the transition to commercial hatchery production was never accomplished. However, in 1972, the Oceanic Institute proposed to the Rockefeller Foundation that a number of species could be mass-produced as cheap food fish for the rural poor if interdisciplinary teams of scientists and technologists tackled the problem. The institute proposed that four regional research and development centers be created worldwide, similar to the famous agricultural centers of the so-called Green Revolution being funded by the Rockefeller Foundation, within a consortium of donors and would be called the Consultative Group for International Agricultural Research.

John Pino, a divisional director of the foundation, understood the financial burden to the international donor community of the many agricultural research centers, such as the International Rice Research Institute in the Philippines and the Vegetable Research Institute in Taiwan. He therefore proposed a more flexible organization with a small core of personnel capable of tackling problems by collaborating with a variety of research, training, and education institutions worldwide, if necessary. Pino sounded out his ideas in Hawaii at a meeting attending by the state's principal marine research institutions. However, the meeting proved to be fruitless, and the organizations failed to show any cohesion for leading such a project. He subsequently asked the Oceanic Institute to prepare an organizational document proposing a single theme and a small institutional core with a simple mandate, which could be refined more closely once the organization was underway. Within a few weeks, a document was drafted by the institute's aquaculture staff in collaboration with Burr Steinbach, the former director of the Marine Laboratory at Woods Hole, and Tim Joyner, the instigator and pioneer of marine farming of Pacific salmon. After some further modifications by Pino and Bill Sellew, a former Wall Street broker and fundraiser for the Oceanic Institute, it was presented to the Rockefeller Foundation's Board of Trustees in May 1974. It was accepted and funded for two years. The new center was called the International Centre for Living Aquatic Resources Management, an awful mouthful of words that quickly encouraged the use of ICLARM, a more easily recalled acronym.

ICLARM's electoral system for governors and officers was modeled after that of the Woods Hole Oceanographic Institution, and the new governing board appointed Phil Helfrich, a tropical fisheries scientist at the Hawaii Institute of Marine Biology, as the first director in 1975. The organization was based in Hawaii, which was acceptable to the Rockefeller Foundation *pro tem,* because the principal targets of its research programs were to be the islands of Oceania. Helfrich was then to negotiate for a permanent headquarters in Fiji. However, it soon became clear that a well-funded research program in small-scale fisheries and aquaculture could make much more impact around the rim of the Pacific Ocean rather than within it, and therefore, it was decided by the board that another home had to be found for the organization. The proposed options were Taiwan, Thailand, Malaysia, the Philippines, Indonesia, and tiny Singapore. The latter was everyone's first choice, not only because of its location at the hub of international aviation, but also because it had no extensive background in aquaculture. Consequently, any of the natural benefits that typically accrue to the host country of an international organization would be very modest in Singapore and should not create envy among the other aquaculture giants—as had occurred some years before with SEAFDEC and its Aquaculture Department in the Philippines. However, a Singapore headquarters for ICLARM was not to be. The quick political and financial acumen of President Marcos soon had yet another center located in the Philippines. The Rockefeller Foundation accepted Marcos's offer to host ICLARM and to accord it the usual international financial privileges. In reality, developing a regional program from the Philippines readily suited the foundation, because it could organize the logistical and financial operations through another of its programs in the country, the International Rice Research Institute, until ICLARM could function independently.

When the option for directing the program from Fiji vanished, Phil Helfrich preferred to remain in Hawaii rather than relocate to the Philippines. He handed the reins of the ICLARM directorship to Jack Marr. Marr was a traditional fisheries management scientist who had led FAO's Indian Ocean Program for many years. He had little time for the growing field of aquaculture. Consequently, apart from John Pino's ideal ICLARM project—to create and manage a culture-based salmon fishery in the Southern Ocean—the early projects of the new organization under Marr were mostly concerned with small-scale fisheries. Anything else to do with aquaculture had to be funded separately.

With the Rockefeller Foundation's confining its financial support mainly to a core budget, project funding continued to dog the survival of ICLARM for the next decade. It was not until 1992 that it was finally accepted under the umbrella of the consultative group, about twenty years after such an idea was first proffered.

11.7   Kyoto and the United Nations Development Programme/Food and Agriculture Organization Global Program

Just after the Oceanic Institute's proposal in 1972 to the Rockefeller Foundation for four world centers for aquaculture, Bill Ripley, the fisheries adviser to UNDP in New York, and Ramu Pillay, a senior fisheries officer of FAO in Rome, began to discuss a plan for the United Nations to fund a number of regional centers for aquaculture research and training. All this would be executed by FAO through a new global program, with Pillay at the helm. However, achieving anything in the United Nations system is a slow process. It requires a logical and formal progression of recorded events, complete with supporting documents and democratic decisions by members who meet once every two years, to justify each subsequent step. Further, the starting point is always an initiative by one or more member countries, because the funding organizations within the United Nations' system can respond only to individual or collective requests from member governments. They cannot be seen to be instigating programs of their own interest. However, this potential constraint can always be circumnavigated with nimble footwork by those who know the tune and the steps.

Ripley and Pillay were two such dancers, both well steeped in the United Nations' choreography. First, they persuaded UNDP to fund a project for a small core of people within FAO to organize a world technical conference on aquaculture in 1976. They argued, quite reasonably and justifiably, that a global conference on aquaculture would focus attention on the importance of this emerging sector for developing countries and would provide some organization for development. Moreover, the timing was right. It would be part of FAO's regular effort to hold global meetings on major fisheries issues, and it had been almost a decade since the important "World Symposium on Warm-Water Pond Fish Culture" in Rome in 1966 (see description, Section 9.1). However, as part of their approach, Ripley and Pillay also obtained preparatory assistance funds for a series of regional planning meetings. These preliminary gatherings would be a forum for the invited regional delegates, who were responsible for aquaculture or inland fisheries in their respective countries, to define and outline their own specific priorities and needs. Hence, they would be the source of the required initiatives from among member countries. The preparatory assistance was sufficient to hold regional planning meetings in Africa, Asia, and South America in the last six months of 1975. The first was in Ghana in July, followed by Thailand in October, and Venezuela in November.

The "Technical Conference on Aquaculture" at Kyoto, Japan, in 1976 was a huge success, due in no small part to the generosity of the host country. It was also a great success for Ripley and Pillay. Perhaps because it was an FAO gathering, the official records show that there were 308 participants and 155 observers, and the national delegations included many of the leading aquaculture researchers of the time, together with a surprisingly large number of individuals from the private sector. Everyone had seized the opportunity to verify claims about aquaculture in Japan and to learn all that they could from their visit. Pillay also increased the appeal of the meeting with a nice original touch. He introduced an array of prizes for films about all aspects of aquaculture, so that delegates had a chance to see what else was happening in the world.

The majority of participants, most of whom were from developed countries, were blissfully unaware of any grand plan behind the Kyoto Conference. As a result, as the meeting drew to a close with the formulation of the conference doctrine for the future, the participants contributed genuine conclusions and recommendations that dealt seriously with global priorities for further development. In particular, they saw the lead being taken by the private sector with products to meet increasing market demand: that is, farming the highly profitable commercial species, such as salmon, catfish, and marine shrimp. At that time, however, the idea of any development project to put money in the pockets of private entrepreneurs was anathema to the United Nations' system. It had to confine its interests to government-led projects to meet national goals of food self-sufficiency and to provide subsistence protein for the rural poor.

Consequently, a small drafting committee had to deal with the ideas and recommendations derived from two separate schools. On the one side were the humanitarians, led by the FAO and delegations from the international agencies, who were eager to have commitments to developing countries. On the other, there were the pragmatists, who knew that only the market would drive growth. This group was led by Peter Hjul, the editor of the magazine, *Fish Farming International*. He was strongly supported by delegates from the commercial sector, eager to put limited financial resources into high-priority payoff. Each evening, the report of the drafting committee was passed through to the secretariat of the conference, mainly made up of FAO personnel, who then proceeded to massage out much of that which was recommended by the pragmatists from the floor. When the report was read back the next morning, the conference broke out in uproar, and the cycle started all over again.

As time ran out, the Kyoto Declaration on Aquaculture had to be something of a compromise between the humanitarians and the pragmatists. No references were made in the declaration to priorities of species, and the broad statements about the benefits of aquaculture were sufficiently ambiguous to fulfill both humanitarian and commercial interests. But the choreography for the work of Ripley was complete, and the declaration had clear grassroots statements on the need for a network of centers in key regions that could be used for research, technical training, and the exchange of latest information in the field. This was sufficient for Ripley's chiefs in New York. The core at FAO was formalized into a new inter-regional project called the Aquaculture Development and

Coordination Programme (ADCP), with Pillay as the program leader. New staff was hired, and the project began the process of creating the regional centers. By the end of the decade, there were four fully staffed centers, operated by FAO and funded by UNDP and other major donors. These were the African Regional Aquaculture Centre at Port Harcourt in Nigeria, the Regional Centre for Aquaculture in Latin America at Pirassununga in Brazil, the Network of Aquaculture Centers in Asia headquartered at Bangkok in Thailand, and the Mediterranean Regional Aquaculture Program with headquarters at Tunis in Tunisia.

11.8   A plethora of experts

The increasing interest in aquaculture by the private sector in the 1970s was due in considerable part to a growing number of consultancy firms. The firms were staffed by a variety of people. There were many American and British volunteers who had returned with practical experience garnered from their tours overseas. There were scientists and laboratory technicians, most of whom had built up experience in national research and development facilities. Finally, there were also some "overnight experts" who had obtained their information from several technical books and papers that were just starting to be published.

In the United States, one of the principal individual consultants was Harold Webber, an engineer with no background or experience in the field. However, he had traveled widely and had a photographic memory for what he saw. With his persuasive tongue and a direct line to many of the largest corporations in the country, he led a group of technical experts responsible for many of the large commercial investments in marine shrimp production in the industry through the 1970s. Some were content to begin production in the southern United States, but he was soon able to persuade them to move to Central and South America, where the temperatures were more conducive to faster growth of the shrimp.

Other consultants were more formal. They were usually structured in large architect and engineering companies and provided design and engineering services for farm sites. Others went further and provided complete services, including technical and financial management, and even marketing. One of the first was KCM International, founded by Ron Mayo in 1978. His background in waste-treatment had enabled the parent company, Kramer, Chin, and Mayo, to become one of the top engineering firms in North America for salmon and sport fish hatcheries and aquariums. This experience made the conversion to aquaculture farms easy. The company did not confine its work to fish, and was responsible for some major projects throughout the world on a variety of species. One of its earliest successes was the design and construction management of the first marine shrimp hatchery in China in 1978. The project in the ice-cold Gulf of Bohai seemed nonsensical to Mayo at the time. The hatchery had to be extraordinarily large, because it had to provide all the juveniles at the same time for a single four-month growout season. But, as was often the case for many early aquaculture development projects, the investors behind the project, in this case from mainland China and Hong Kong, had hidden agendas that were far more profitable than raising shrimp.

Despite the excitement of developing large aquaculture projects in foreign countries, the effort was not without its difficulties. KCM International followed its success in China by winning a large contract to design a fisheries research and training center, a carp hatchery, and a model homestead farm complex at El Abassa, Egypt. The location was an extensive low-lying marshland in the east of the Nile Delta, where the opulent King Farouk built himself a shooting lodge in the 1950s. In addition to the abundant resident and migrating wildfowl, the area was a haven for malaria and schistosomiasis. The two-phase project was funded by the U.S. Agency for International Development as part of the economic development package given to the country after Egyptian President Sadat and Israeli Prime Minister Begin signed the Camp David Peace Treaty in 1978. Unfortunately, the competition between American consulting companies and their Egyptian partners for the second phase of the project that addressed education and training became a very contentious issue, and subsequently a number of people landed in the Cairo jail.

Another company that contributed significantly to aquaculture growth was Aquatic Farms. This company had been started in Hawaii by Ed Scura at about the same time that KCM International was incorporated. Scura's formula for success was a company investment in a fully operational farm for producing freshwater prawns. This clearly demonstrated to his clients the hands-on experience and full service capabilities of his small group of consultants. Subsequently, when the global tide turned from freshwater prawn farming to the more successful and profitable marine shrimp farming, he converted his farm accordingly. This move enabled Aquatic Farms to win several of the large contracts being awarded by the Asian Development Bank, and later the World Bank, for marine shrimp farming development projects in Indonesia, the Philippines, and India.

In Europe, many of the early consultancy groups were not totally dependent on their endeavors for survival, as were their competitors in the United States. The most professional company in Europe was France Aquaculture, which was a quasi-government organization providing full services to its clients. The financial operations of the *société*, as the organization was structured, were made very simple and profitable. Jacques Perrot, the first director general, was able to draw on technical experts, such as Michel Girin, Alain Michel, and Jean-Michel Griessinger when required, from the Centre National pour l'Exploitation des Océans (later, the Institut Français pour l'Exploitation de la Mer). The government's marine science research institution just happened to be an investor in the *société* and had voting representation on its board. With the large resources of such an organization behind it, France Aquaculture could also offer hands-on training at over twenty-four government-run research facilities scattered throughout France and the French territories, such as the Centre Océanologique du Pacifique in Tahiti, and in New Caledonia, Martinique, Guadaloupe, and French Guayana. Furthermore, through its overseas embassies, the French government frequently provided grant funding for project feasibility studies, particularly in the old French colonies, and scholarships for associated project training and education.

With such financial advantages, France Aquaculture was a major competitor for any project open to international consulting companies. However, to its great credit, France Aquaculture was the only consulting company that would categorically guarantee success for the investors. If its hatcheries did not produce the required number of juveniles on schedule, then France Aquaculture was under contract to compensate the investor and put the matter right at its expense. On the other hand, if the hatcheries met their targets for the first five years, then the investor would pay previously agreed, annual royalties. These guarantees automatically made the company more expensive to hire than other competitors, but without doubt, France Aquaculture was responsible for introducing more countries and companies to aquaculture than was any other consulting company in the world. Regrettably, the success story had an unhappy ending. As a semigovernment entity, it was not entirely in control of its destiny, and in 1994, it was sold for a peppercorn to Cofrepêche, another quasi-government *société* that was struggling to make ends meet in fisheries consulting. For a time, the organization operated under a joint name, but in 1997, after its coffers were quietly sucked dry, the name of France Aquaculture was ignominiously obliterated, and the remaining aquaculture experts were muscled out.

Another highly effective international consulting group was Agroinvest from Hungary, which was established in 1979. Agroinvest was the export arm of an internal parastatal body called Agrober, which was created in the organizational structure of the government's Department of Fisheries to introduce research achievements of the Fish Culture Research Institute at Szarvas. Consequently, like France Aquaculture, Agroinvest was able to call on publicly paid technical staff at the Fish Culture Research Institute, where the company was also based. Therefore, the consulting staff of Agroinvest was particularly strong at that time in the major freshwater fish species, such as the carps, together with polyculture practices and integrated farming with ducks. The staff included experts such as Elek Woynarovich, Andras Nàgy, Janos Bakos, Lazlo Varadi, and Janos Oláh. Long past an age when he might have retired, Woynarovich could always be found in some remote part of the world, devoting long periods of his life helping a development project to raise fish with pigs and poultry.

A similar level of cooperation by the government to a quasi-private sector was also evident in Israel. The growth of aquaculture in the country was due largely to the independent *kibbutzim* system, in which each *kibbutz* sought profitability through the combined talents of its residents. Consequently, certain *kibbutzim* noted for their highly productive fish farms were pleased to offer the international consulting services of their residents—experts such as Shmuel Sarig, Shimon Tal, Balfour Hepher, and Dan Mires, many of whom worked for the Ministry of Agriculture or for a university. In addition, they were able to draw on other scientific and technical researchers still working at the national Fish and Aquaculture Stations at Dor and Eilat. These included key figures such as Rom Moav, Hillel Gordon, and George Kissel.

However, the death of France Aquaculture in 1997 was essentially the end of the full-service consulting companies. Competition from new farming technology and hatchery engineering companies in highly-active Asian countries, such as Taiwan and the Philippines, began to take over—and rightly so. The regional firms cost the clients far less than did the Western companies, which had the burden of higher salaries and expensive travel, and they employed experts who had practical in-country experience. Further, the regional firms, and particularly any owned and operated by a woman, such as Elvira Baluyut of Primex, Inc., in the Philippines, became priority choices of the World Bank and its regional development banks to manage assistance projects. This new selection policy effectively ended the lucrative international contracts on which the full-service Western companies had built up their businesses for over a decade, leaving them to chase the dwindling number of fisheries and aquaculture projects created by their own national bilateral agencies.

11.9   The printed word

The practical but frequently narrow experience of the growing number of independent consultants looking to make a living selling information to the young aquaculture industry in the early 1970s was broadened substantially by the sudden publication of industrial and trade news for the growing number of small producers. Through the emerging years of aquaculture in the 1960s, the only information freely available to potential investors was the growing number of published scientific papers and technical articles written by researchers in publicly operated agencies. But not all researchers and research organizations were free with their information at first. The large private corporations kept their commissioned studies strictly confidential, and some parastatal or quasi-government organizations, such as the White Fish Authority in Britain, kept their research and development results on fish and oyster farming carefully under wraps for the benefit, first, of any entrepreneur in the national fishing industry.

In Europe, this yawning vacuum for farm management information was filled in 1973 by the first publication of *Fish Farming International*, by Arthur Heighway Publications Ltd. in England. Heighway had taken over the reins of the ailing *Fishing News* in 1953. Almost immediately, he forged close links to the new FAO organization in Rome, which was clearly becoming a great source of fisheries information—and it was all publicly available. Using material from FAO conferences and meetings, he began a series he called Fishing News Books. Early in the 1970s, with international interest growing in aquaculture, FAO started to produce a quarterly magazine called the *FAO Aquaculture Bulletin*. It was a high-quality glossy publication, but it dealt mostly with projects in developing countries. The issues were published erratically, and it was costly for FAO to produce and distribute it free of charge. Ramu Pillay, who was with FAO in Rome preparing the groundwork for the UNDP Global Program with Bill Ripley, urged Heighway, at that time, the chairman of *Fishing News*, to consider taking aquaculture information out of his two existing trade publications, *Fishing News* and *Fishing News International*, and put it into a new magazine or paper devoted solely to aquaculture.

The editor of *Fishing News* was Peter Hjul, a journalist who had been associated with trade newspapers all his working life. As a young reporter in the docklands and railyards of Cape Province, he was always more at ease with the fishermen, and he began to specialize in the fishing industry of South Africa. But he was also an antiapartheid activist, and after two years of being under virtual house arrest, he and his family fled for their lives to England in 1965. He joined the staff of *Fishing News* and soon got on well with Heighway. Subsequently, the two of them decided to try Pillay's idea to add an aquaculture paper to their portfolio, with Hjul as editor.

Fish Farming International was launched in 1973 as a small, almost pocketbook-style, glossy publication. Advertisements were few at first, and publication through 1974 was sporadic. In 1975, the format was changed to that of a quarterly magazine. With regular issues, and the active involvement of Hjul in the grass roots of the industry, such as sponsoring conferences and organizing trade exhibitions, the advertising increased. *Fish Farming International* (FFI, as it was to be fondly called) was on its way. In 1984, the magazine became a monthly tabloid-format newspaper, sometimes filling sixty-four pages of technical information, general articles, reviews, announcements, and advertisements. Here was something fish and shellfish producers all over the world wanted to read, and Hjul stayed at the helm of the paper for almost twenty-five years.

Another trade journal that followed close on the heels of *Fish Farming International* was *Fish Farmer*, first published as a quarterly in 1977, and later as a bimonthly. The magazine was specifically directed to fish farmers and not to the impersonal face of the fish farming industry at large. It was published by the I.P.C. Business Press as a complement to its extraordinarily popular *Farmers Weekly*, a paper that could be found on the table of almost every farmhouse in the British Isles. Because of the small numbers of actual fish farmers at that time, its circulation as an expensively produced glossy magazine was somewhat limited, and it was not until Stuart Barnes took over as editor in 1983 that it really came of age. Barnes subsequently bought the paper through Amber Publications in 1986. Without losing sight of the magazine's original goal and its intended readership, he diversified its content and provided a special international section that significantly raised its popularity worldwide.

A more robust scenario was taking place in the prospective aquaculture publishing industry in the United States. The catfish industry, which had suddenly burst on the scene and surpassed the traditional trout industry, was served in the late 1960s by a tabloid, *The Catfish Journal*, which later gave way to the *Catfish Farmer*, produced by Porter Briggs and Bill Glasscock in Littlerock, Arkansas. There was also *Fish Farming Industries*, a magazine published in Chicago. Then, in December 1969, yet another independent magazine was launched. It was called *American Fish Farmer and World Aquaculture News*, and it was targeted at both North and Central America. Therefore, the competition for the small market at the time was very intense. Eventually, Briggs bought out the opposition in 1973 and combined all three into one new magazine called *The Commercial Fish Farmer and Aquaculture News*. In time, this became the *Aquaculture Magazine*. Although the magazine was later sold to a company in North Carolina, Maurice Moore was the driving force behind most of the aquaculture news that came through those pages, and he remained editor of one or more of these popular fish farming magazines for twenty years.

Scientific and technical journals devoted to both freshwater and marine fisheries had been around since the end of the nineteenth century, and they catered to the new breed of fish culturists, and published their first papers. These were mainly the house journals of the early societies, such as the *Transactions of the American Fisheries Society*, first published in 1872; the *Journal of the Marine Biological Association of the United Kingdom*; and *Journal du Conseil*. The first journal truly devoted to fish farmers was probably *Ceskoslovensky rybar* (The Bohemian-Moravian Fisherman). It was founded and edited in 1901 by F. Vesely. Subsequently, after nationalization in 1918, the journal was taken over by the Central Union of Fishermen and edited for many years by V.J. Stepan, the director of the Fishery School at Vodñany.

The next major landmark year in aquaculture publishing was 1934, with the appearance of the *Progressive Fish Culturist*. It was published by the U.S. Bureau of Fisheries, despite the general austerity of the country at the time. But the pocket-sized journal proved to be a singular success, with its highly practical information on any aspect of fish culture in a simple, low-key presentation. It would stand the test of time for the next sixty-five years before its name was changed in 1999 under its new ownership by the American Fisheries Society.

It was not until after the Second World War that new scientific and technical publications about fish culture began to appear again. *Bamidgeh* was the first, issued as a quarterly in 1955 and edited by Shmuel Sarig at the Nir David Research Laboratory in Israel. It was published jointly by the Department of Fisheries and the Fish Breeders Association. *Bamidgeh* had actually started life in 1949, just after national independence, as a monthly mimeographed bulletin written in both Hebrew and English for the new immigrant *kibbutniks* eager to produce fish. *Bamidgeh* was followed in 1971 by *Prace VURH Vodñany*, the published work of the Institute for Fisheries and Hydrobiology in Vodñany, Czechoslovakia. Both of these publications were essentially national journals. The first truly international journal devoted entirely to aquaculture was in fact called *Aquaculture*, and it was published in Amsterdam by Elsevier Science B.V. The first issue appeared in 1971, with Canadian Don Alderdice and Dutchman Bas De Groot as editors.

Hard-backed books about aquaculture soon followed the scientific and technical journals. Because most of the fish production in the Western Hemisphere centered around the salmonids, several practical manuals had been published in the United States by the U.S. Fish and Wildlife Service through the 1950s and 1960s. After a prolonged effort to integrate them into one volume, the result was a book called *Fish Hatchery Management*, edited by Robert Piper. It contained such a large amount of detail about all aspects of fish production and hatcheries that it proved to be invaluable for culturists raising many other freshwater and marine species. The book had to be reissued and revised several times over the years. Another book based on salmonids that also had wider importance for the engineering of aquaculture facilities was *Design of Fishways*, by C.H. Clay.

The first book specifically dedicated to the new field had the simple title, *Aquaculture*. This large work, first published in the United States in 1972, was essentially the synthesis of material collected by John Bardach and John Ryther as they made their fact-finding world tour for the Stratton Commission in the late 1960s. The book was very detailed and had few illustrations, and its market was restricted at first to those already in the field. In contrast, *Farming the Edge of the Sea*, written by Ed Iversen and published in 1976, was packed full of photographs. Its text was more widely appealing, and its timely title helped it find a strong market beyond the scientific and technical specialists.

Both of these books, however, were only good overviews of the aquaculture field at the end of the 1960s. Other books that subsequently started to appear on the shelves were much more specific to individual would-be culturists. These included, for example, *Catfish Farming Handbook* by Jerry Mack (1971), *Fish and Shellfish Farming in Coastal Waters* by Peter Milne (1972), *Culture of Bivalve Molluscs* by Peter Walne (1974), *Eel Culture* by Atsushi Usui (1974), and *Shrimp Culture in Japan* by Kunihiko Shigueno (1975). All these books, two of which were published by Arthur Heighway under his newly formed banner of Fishing News Books, were the first to present a detailed introduction to the problems of farm production by genera or species, and to the engineering of farm facilities and hatcheries. All of them proved to be very practical guidebooks throughout the decade of the 1970s for the new wave of fish and shellfish culturists beginning to swell the ranks.

Observing this new technology and its growing market for information, the large publishing houses soon stepped in. Elsevier Science Publishers of Amsterdam once more led the way and started its series, *Development in Aquaculture and Fisheries Science*. The first four volumes, all published in 1976, were written by Peter Korringa, director of the Institute for Fisheries Investigations in the Netherlands, who had studied mollusk farming since 1937 and made innumerable reports about his trips. Consequently, his books described in very practical and financial detail, farm enterprises in seaweeds, mussel, oyster, clams, and shrimp production that he had personally collected on working visits around the world.

It would be another decade before significant works appeared on aquaculture in Asia. For the most part, eager authors and editors in the third world countries had to wait for the financial support of the bilateral assistance agencies, such as the International Development Research Centre of Canada or the International Center for Living Aquatic Resources Management, to translate and publish their research or their wisdom. Others sought the help of Western ghost writers—some of whom preferred not to remain too hidden and took first authorship.

11.10   Aquaculture insurance

In addition to the contributions of the consulting companies and the publishing world, both of which had major impacts on the growth of the global aquaculture industry through the 1970s, another industry was beginning to make its presence felt. The business was livestock insurance. The pioneer behind its presence and ideas was Paddy Secretan, an insurance specialist with substantial family background in the business at Lloyds of London. His experience told him that without the influence and benefit of insurance, producers would continue to repeat the same mistakes over and over again in the design and operation of their farms. Consequently, he raised capital to establish Aquaculture Insurance Services, a company that was prepared to place livestock insurance with Lloyds.

Like many others who realized the importance of the FAO Technical Conference in Kyoto in 1976, Secretan paid his own way to Japan to speak. His ideas were completely revolutionary, and only a few of the producers who were there understood their value. For the most part, there was general skepticism that the livestock insurance business could be profitable. But Secretan knew he was right, and he persisted. Slowly, his insurance business grew. He was careful in selecting his clients, because his early experience in brokering insurance and adjusting claims had given him a wealth of knowledge in identifying operations that were exposed to the greatest risks. This experience he passed on to producers through the terms and conditions of their policies, by his ceaseless attendance at trade shows, and by his characteristically forthright papers at technical conferences. Later, he operated his own series of risk-management conferences.

Almost single-handedly, Secretan was responsible for developing higher standards of fish farming practices, particularly throughout Europe and Scandinavia, but also in the United States. Through the 1980s, the insurance business in Europe and North America expanded rapidly with the increasing number of Atlantic salmon, Pacific salmon, and catfish farms. New insurance brokers entered the field, and the competition grew. Unfortunately, in the fight for business, many of the insurance companies issued policies that were ill-advised. Field agents were unfamiliar with the special problems associated with fish and shellfish farming, and their inexperience soon showed in the policies they wrote. The standards Paddy Secretan developed from his years of experience were frequently forgotten. Much to the chagrin of the underwriters, many large settlements would be paid out in years to come on losses due to the perennial risks from algal blooms and poorly designed facilities.

Bibliography

Asui, A. (1974) *Eel Culture*. Fishing News Books, Farnham, Surrey, England.

Clay, C.H. (1995) *Design of Fishways and other Fish Facilities*. Lewis Publishers, CRC Press, Boca Raton, FL.

International Foundation for Science (1980) Proceedings of First International Meeting, 1978, Universiti Sains Malaysia, Penang. *Aquaculture*. 20 (3), by Elsevier Science.

Iverson, E. (1976) *Farming the Edge of the Sea*. Fishing News Books, Ltd., Surrey, England.

Korringa, P. (1976) *Development in Aquaculture and Fisheries Science*, 4 vols. Elsevier Science Publishers, Amsterdam, The Netherlands.

Mack, J. (1971) *Catfish Farming Handbook*. Educator Books, San Angelo, TX.

Piper, R. (1982) *Fish Hatchery Management*. United States Department of the Interior, Fish and Wildlife Service, Washington, D.C.

Shigueno, K. (1975) *Shrimp Culture in Japan*. Association for International Technical Promotion, Japan.

Walne, P.R. (1974) *Culture of Bivalve Molluscs*. John Wiley & Sons, Inc., New York.

Chapter 12

Building Global Capacity (1980–2000)

Abstract

The 1980s and 1990s witnessed great interest in aquaculture, with expansion in growth and diversity. With stability, the industry gained professionalism. For investors, farming was controlled by financially minded managers. These decades represented a search for economies of scale and more species. Vertically integrated operations required economies of scale for farms and other cost centers, hatcheries, and marketing outlets, and saw floating salmon pens growing to thirty-thousand-cubic-meter sea cages. Terrestrial shrimp farms grew to one hundred hectares, shrinking back to two-hectare by the 1990s. Funding grew from World Bank and individual countries; the 1999 Japanese agricultural fisheries budget was $1 billion—half government, half private. Statistics improved after 1983; the Fishery Information, Data, and Statistics Service of United Nations Food and Agriculture Organization formed and grew. The last decade saw large numbers of new species under commercial culture in demand in national or international markets.

12.1 Progress and problems of the eighties

The 1980s witnessed great interest in the new field of aquaculture, and the decade was one of remarkable expansion in growth and diversity. Although the long-term trend was at that time clearly upward, the ascent was still a succession of short-term peaks and valleys. Real obstacles to production of one species or another continued to appear, but they were relatively small compared with those that had blocked all progress in decades gone by. They were stumbling blocks and

---

The History of Aquaculture. By C. E. Nash. Published 2011 by Blackwell Publishing Ltd.

not dangerous impediments to growth. One by one, as production technologies became more reliable, the numbers of producers quickly surpassed the critical mass necessary to stabilize their individual corners of the industry. Here at last was the real evidence that aquaculture was being built on substance and not still the wild dreams of a blue revolution in marine science and technology that had plagued the previous two decades.

Nonetheless, the realities "down on the farm" were still frustrating many would-be producers. Their financial figures were not always so encouraging. Many biological successes in the laboratory, when scaled up to pilot-scale levels, still revealed deficiencies in biological knowledge and production technology. Even some successful pilot-scale projects simply failed to make the grade economically at a commercial scale. Although these problems of transition were costly to the industry in the early 1980s, they proved to be very important. For almost two decades, the emerging field of aquaculture had been dominated by industrious and practical-minded biologists, many of whom, through trial and error at someone else's expense, had become self-made jacks-of-all-trades. As it gained stability, the industry opened up a new field for the professionals, and it was immediately attractive and interesting employment for mechanical engineers, agricultural engineers, food technologists, veterinarians, and economists. On behalf of the investors, the business of operating a farm was controlled by financially minded managers.

As the industry began to stand on its own feet and to spend its own capital, decisions and changes came hard and fast. Many once-promising species, such as plaice, common sole, flounder, mullet, pompano, and rabbitfish, were dropped, because propagation of the numbers required for growout continued to be unreliable. There were more unexpected problems with compounded artificial feeds, which often proved to be nutritionally deficient. Further, the logistics of manufacturing and delivering increasingly large quantities had not been adequately estimated. Large losses of stock were encountered through poor nutrition, exposing the animals to a variety of new diseases for which there were no known treatments. There were many unforeseen disasters. In addition to the normal ravages of storms and floods, entire farm populations were wiped out by toxic algal blooms, or water pollution, or general failure of mechanical equipment.

For many farmers working with newer marine species, the early 1980s became a period of intensive adaptive research, particularly backtracking on nutrition, pathology, and engineering, while the scientists explored new directions in induced breeding and genetics. For them, getting any return on investment was difficult, because the sporadic quantity and quality of farmed products could not compete on the same markets with natural wild-caught products. In contrast, farmers producing the more traditional freshwater species that had existing markets, such as Chinese and Indian carps, salmon and trout, catfish, and tilapia, began to flourish. The one exception was the producers of marine shrimp, and many thousands flocked to invest in farms wherever it was thought that marine shrimp would grow.

Fortunately, the 1980s was a period when investment capital and research funding was readily accessible. In general, the global economy was beginning to boom, and particularly in the majority of Asian countries, which were already steeped in the traditions of aquaculture and which had increasing populations that demanded fish. In Europe and the Americas, where aquaculture was still a relatively unknown but exciting prospect, national funding agencies readily poured in money. Governments provided support for both fundamental and applied research, and there was considerable private investment by large corporations already working in animal nutrition, pathology, and engineering. In the developing countries, there was little government support for fundamental research, but the international assistance agencies carried the load, mostly financing only applied research. This was no mean contribution. By the end of the decade, for example, the United Nations system, Trust Funds, and the bilateral aid agencies had contributed over $300 million to development projects, primarily in secondary institutional support, such as research, education, training, and infrastructure for these activities.

Industrial investment was carried almost entirely by the private sector. In the developed countries, this was often assisted by government incentives, such as subsidies, reduced taxation, and exemption from certain duties. Members of the European Economic Community (EEC) and its affiliated groups of commonwealth and former colonial countries received very generous development grants for construction, as well as special tariffs on the products.

Although the extent of the private investment in aquaculture through the 1980s is not identifiable, some idea of the scale of investment in the developed world can be obtained by comparison with the investment made by the World Bank system in developing countries. The big attraction was the profitable shrimp farming industry, particularly in Asian countries, but also in countries of Central and South America, and Africa. The World Bank's annual support was well over $1 billion to the industry, and by 1992, was rapidly approaching $2 billion. Although financial support through country loans began as capital investment, the World Bank then announced that it was prepared to support extension schemes, research, training, and technology development. China and India became the largest recipients, with some individual projects that collected between $400 million and $700 million.

The other giant on the financial scene was the European Union. Between 1989 and the end of 1999, the European Union ploughed 16 billion European currency units (ECU) of structural funds into "zones dependent on fisheries and marine fish culture" within its membership of thirteen countries. The funds went directly into specific plans for fisheries and aquaculture, such as fishing vessels, aquaculture farms, hatcheries, processing plants, etc., or into nonspecific plans, such as general industrial and social development to help coastal regions lagging behind European norms. With this immense subsidy, the thirteen countries doubled their aquaculture output from some 620 thousand tonnes in 1986 to 1.2 million tonnes by 1996.

12.2 Bigger is better?

Just like investors in many other industrial technologies, investors in aquaculture production operations throughout the 1980s and early 1990s continued to search for economies of scale. For the small businesses, there was no difficulty. A sixteen-*feddan* carp and tilapia farm in Egypt, or a forty-tonne trout farm in Turkey were known to be the minimum economic units for one man and his family to make a comfortable living. Similarly, in the Scandinavian countries, the profitability of coastal salmon farms was well calculated so that their size and location were restricted in line with the governments' socioeconomic development policies for remote coastal areas. However, when production was left to commercially minded entrepreneurs, some very large numbers began to fly when the economic earnings of farm businesses were projected to meet directors' fees, managerial salaries, and dividends for shareholders. Furthermore, if the farm businesses were to be part of vertically integrated operations, then investors required economies of scale not only for the farms, but for the farms in association with other possible cost centers, such as hatcheries, feed production plants, processing plants, and even marketing outlets. Consequently, in the last twenty years of the century, there were significant increments in the size of hatcheries and farming operations, as investors began to scoop up large lengths of the so-called value-chain to make their projected bottom lines thick and black.

The search for economies of scale in the 1980s was both an advantage and a disadvantage for the industry at the time. It was one factor, for example, behind many technological advances in aquaculture engineering, particularly in the development of net-pen or cage technology for coastal fish farms. The simple floating box structure, which twenty years before had been cobbled together from oil drums, welded walkways, and old fishing nets, gave way to some giant floating or submerged complexes based on galvanized metal collars, rigid net walls, and heavy anchors all designed by computer. The popular net-pens rapidly increased from a modest one thousand to thirty thousand cubic meters in the pursuit of greater fish production in the minimum of space.

The majority of floating cage manufacturers continued to provide larger and larger structures to accommodate the wishes of their clients. However, two pioneers of the next generation of sea cages started the other way around. In 1978, Gary Loverich and Tom Croker started a successful fishing net manufacturing business called Net Systems. The operation was located on Bainbridge Island, not far from the site where Joyner and his team at Manchester had pioneered net-pen farming in the United States. Soon the two of them were dabbling in nets for floating fish pens. When they sold Net Systems in 1995, they turned their attention entirely to aquaculture engineering and created another company called OceanSpar Technologies. Aided by all the latest design engineering technology, they first developed the OceanSpar Sea Cage System, which was based on the concept of floating spars to buoy an enclosure from each corner and to maintain rigidity of the net walls. This design was quickly followed by the Sea Station for more exposed sites. The Sea Station had a central spar and a circular rim, and was designed to minimize the impact of the more powerful offshore waves. These structures were capable of holding over two thousand tonnes of marketable fish.

Perhaps the most advanced offshore complex was constructed by the Bridge-stone Engineered Products Company in Japan. Basically, the farm was a modern offshore platform, which instead of servicing oil-drilling operations, serviced a complex of large floating cages containing yellowtail and other species of marine fish popular on the national markets. The system involved high technology, and operations were monitored electronically from a central office on the platform.

Developments like the Sea Station and the Bridgstone platform increased the opportunities to practice fish farming further and further offshore, beyond the coastal zone, where there was the hazard of possible environmental impact. Escape, however, was not possible for the shellfish farming that was also beginning to thrive in the coastal zones.

The growth of marine shrimp and freshwater prawn farms followed a different path from that of fish farms. The size of a pilot-scale production pond for both shrimp and prawns started at around one hectare or less. However, because small ponds were costly to construct, manageability gave way to reduced capital investment with favorable production results. The size of ponds soared, and many were built with water-surface areas of fifty and one hundred hectares, and even more. But poor water circulation and lack of management control in these enormous ponds soon brought the investors back in line. For the most part, by the end of the 1980s, the optimum water-surface area for a marine shrimp pond settled back to about one to two hectares, and that for freshwater prawns, something less.

Accordingly, the question facing the investor and entrepreneur concerned the number of ponds to build to maximize the investment in the water intake and exhaust systems. The answer was "as many as one could afford," which led to enormous farms and farm complexes. The multilateral financial organizations became the leaders in the rush to build big facilities, because their prime interest was in the numbers. Every project had to represent a very large, multimillion dollar country loan; otherwise, the loan would not be made. In addition, each project had to be justified with a very large number of beneficiaries. One project developed by the Asian Development Bank in Sumatra, Java, and Sulawesi in Indonesia in 1983 planned for conversion of about sixty thousand hectares of existing *tambaks* into intensive shrimp ponds, and another twelve thousand hectares of new pond construction. The World Bank made similar giant inroads into the flat coastal lands of India from Orissa down to Kerala. Unfortunately, the success of the farmers encouraged much parallel "slash and burn" development by speculators, which created such environmental concern in Tamil Nadu, Andra Pradesh, and Pondicherry that a Supreme Court judge in May 1995 directed the three states to restrict all coastal aquaculture within five hundred meters of the high tide line. The directive sent shock-waves throughout all the coastal states of India from Gujarat to West Bengal. All further investment and development in the marine shrimp industry was suspended in India for two years, losing the country a considerable amount of hard currency.

Construction of some of the giant coastal complexes was not financed through external assistance, but rather, by sources within a country. For example, Iran, which was a late entry into marine shrimp farming in the 1990s, developed six farm complexes with a combined area of about eighteen thousand hectares. Each site was constructed with the infrastructure for central life-support systems, and then farmers could buy individual farm units of fifteen to twenty hectares.

Not every country focused its attention on giant, high-density, monoculture schemes. Others were quietly taking another track. Because of the close integration between fishing and aquaculture in Japan, coastal fishery production in the last decade of the millennium increased 24% by volume. It was achieved by a combination of compatible activities, specifically by promoting stock enhancement and sensible use of all resources. The concept was called "agricultural fisheries." It was designed to enhance productivity by restoring and improving the fishing grounds through farming and marine ranching, using all the technological advances developed by the Japanese scientists and engineers. In any one location, implementation of the concept could have included six or seven different activities, such as, the construction of artificial reefs for the protection of released fish and natural propagation; development of suitable habitats for enhancement; creation or restoration of seaweed beds; tideland recovery by dredging, creating waterways, or removing heavy sediments; installing upwelling flow generators to maintain water quality; and using all available production structures. The budget for 1999 was $1 billion, of which the government provided just more than half. The rest was contributed by the private sector and the fishermen's cooperatives.

A similar concept, but without the elaboration, was also being developed in Italy. On the basis of their knowledge of the management of the old *valli* systems, which were characteristic of Italian aquaculture in the thirteenth and fourteenth centuries, Stefano Cataudella and Enzo Vitale decided to try the same approach in the modern setting. Cataudella was a professor at the new second University of Rome, and Vitale was an entrepreneur. The site they chose was the estuary of the River Agri in southern Italy, and they developed their system by closing and connecting the loops and channels of the river's old delta. With the addition of pumped sea water, the result was a farm consisting of irregularly shaped enclosures and canals, each containing water of graded salinity. The ponds were a natural paradise for the traditional Mediterranean species of gray mullet, sea bream, sea bass, and eels, and hosted large populations of resident and migrating waterfowl. The principles behind the operations of Ittica Valdagri, as the farm was called, were a modest but regular quantity of high-quality fish for European markets, indistinguishable from wild fish, and guaranteed to be free from chemicals or unnecessary additives. Cataudella went on to apply these same ecological principles to producing juveniles, replacing the traditional reliance on investment and operation of a hatchery. Again the system worked, and he was able to produce great numbers of juveniles that looked and behaved identically to their natural-born wild cousins.

In keeping with this open-minded attitude toward others, Enzo Vitale opened his farm as a site for training young farmers sponsored by the Mediterranean Regional Aquaculture Programme, which was headquartered in Tunisia. Between 1983 and 1990, the farm played host to several hundred trainees from countries all round the Mediterranean and from as far away as Latin America.

12.3 Numbers cannot lie

At the same time that the global capacity of aquacultue was being built on large investment capital all over the world, it was also being built on the stolid and starchy foundations of statistics. Thanks to FAO, aquaculture statistics suddenly began to appear with increasing regularity.

In preparation for the FAO Technical Conference on Aquaculture at Kyoto in 1976, Ramu Pillay needed some idea of the then-current statistics on aquaculture production to set the stage for the meeting. Because there was almost no information in print, the data had to be assembled through personal contact. Fortunately, Pillay had the services of Michel Vincke and André Coche, two long-time FAO staff members who were trained in classical biology and aquaculture by De Bont and Huet in Belgium and who already had much experience of working in the country's overseas territories. With that background and long list of personal contacts, they spent weeks on the telephone coaxing key individuals to estimate national figures. The response was enormous, but much of the information was still thin, and in the end, many of the national production figures were calculated as theoretical yields based on known areas under production by species. By the time the conference opened on May 26, Vincke and Coche had identified some sixty seven countries as producers of the principal aquaculture commodities (fish, crustaceans, mollusks, and seaweeds), and world production stood at the grand total of just over six million tonnes.

In the years immediately after the Kyoto Conference, which gave birth to gave birth to ADCP under the leadership of Pillay in Rome, the group within FAO continued to maintain a handle on statistical growth. But the reports were intermittent. The next figure for global production, 8.7 million tonnes, came out for 1980, and there was another gap of three years before the 1983 figure of 10.2 million tonnes was announced.

By then, it was clear that the responsibility of gathering and publishing the data had become too much and too costly for ADCP to continue. The Fishery Information, Data, and Statistics Service (FIDI) of FAO immediately stepped up to take over the operation and to collect and collate the figures on an annual basis. Adapting the classification system it used for capture fisheries with groups of similar families and genera, and using its direct links to the fisheries departments of all FAO-member nations, FIDI gathered and published the first book of production statistics for the year 1984. The estimate for global production by the professionals in FIDI that year was 10.07 million tonnes, a figure very close to that calculated by the amateur sleuths in ADCP and their network of friends.

Statistics compiled and published by FAO are usually accepted as being fairly accurate, and therefore its figures were sufficient to silence most of the remaining aquaculture skeptics. Nonetheless, gross production figures of fish and shellfish in millions of tonnes were still not easy to interpret by those not necessarily knowledgeable about the fisheries field. For most people, fish have to be dressed out; heads of crustaceans are removed; mollusks are shucked from their shells; and seaweeds are cleaned and dried. So, what did the figures all mean to the man in the street? An answer was never forthcoming. Because of its vast databank of information on fish prices and trade, FIDI was able to estimate gross production of aquaculture commodities in terms of their financial value. Gross production in 1984 had a reported value of over $12.5 billion. That, at last, was something everybody could understand.

With FIDI firmly in control, production and value figures were published annually from then on. By the end of the decade (1989), production had reached 14.02 million tonnes and had a value of $24.6 billion. Of course, not all the increases were necessarily real growth at first. Many new countries began to report data for the first time, and several others had to go back and begin to separate more clearly the true aquaculture data from the natural harvests of inland waters and coastal mollusk beds that had been managed for decades. Nonetheless, the growth and financial figures for global aquaculture by the end of the 1980s were impressive, and they were made more significant by the relative stagnation of the global harvest from capture fisheries. With the authority of FAO statistics behind it, the aquaculture industry began to catch the serious attention of government administrators and bankers.

12.4   More species under culture

Although no detailed information on species was available in the 1975 census made for the FAO Technical Conference in Kyoto, priorities for highly marketable species became increasingly evident in data provided by FAO each year since 1984. By the end of the millennium, FAO statistical data recorded that over 150 aquatic species were cultured in some form or another, comprising finfish (39 species), crustaceans (23), mollusks (35), algae (4), and miscellaneous aquatic animals, such as frogs, turtles, sea-squirts, pearl oysters, and sponges. Of these, some 34 were obviously the principal marketable species farmed in many countries over a wide geographic area, and their individual total production was more than twenty thousand tonnes annually.

The freshwater fishes, which were easily farmed and required few costly inputs, contributed the most to global production. Key species remained those that were of traditional value and had been in demand at any price in most Asian and some East European countries for centuries, particularly silver carp, bighead carp, and common carp. Other species of finfish that were widely used in pond culture in Southeast Asia included Siamese gourami and Java barb. The Indian carps also continued to be traditional and important in West Asian countries, but their substantial production had not yet been well identified.

The tilapias continued to have importance for small-scale farming to suit local village markets in Africa, Asia, and Latin America. The Nile tilapia was the most commercial of those low-value species. However, many of its hybrids, particularly red tilapia, had demonstrated appeal to small, medium-value markets, and were better for small-scale entrepreneurial farming not only in these traditional regions, but more recently on the supermarket shelves of North America and Europe. Similarly, the diverse family of catfishes was justifying its considerable

Figure 12.1 Iran, 2000; sturgeon collage: feed-drying machine (upper left); feed packaging (lower left); young sturgeon in hatchery (upper right); recaptured adult sturgeon (lower right).

medium-value potential. Both local and institutional markets had been developed in the last decade for these higher-priced freshwater species, particularly channel catfish in the Americas.

Of the fishes that tolerated both fresh and marine waters, many of which had been of significant importance worldwide for a century or more, high-value sturgeon were always valued for their jewelry as well as their nutrition, but the salmonids remained at the forefront of demand, and accordingly, of production. Pan-sized rainbow trout was the most widely cultured for both retail and restaurant trades, but the decade had seen explosive production of Atlantic salmon in most temperate countries in both the Northern and Southern Hemispheres. Because of oversupply, many product forms had been made available, and farm salmonids were sold dressed, in various sizes, in portions, and smoked. There was a reversal seen in the production of milkfish. Because this was a relatively low-value, staple fish, traditionally used in many countries in Asia, demand had not fallen, yet production and interest in milkfish had decreased as coastal ponds built originally for milkfish farming had been converted for production of high-value marine shrimp in schemes financed by the international development banks.

In general, marine fish were high in demand on local markets and active in international trade. Compared with freshwater fish, they were medium-high in value, but total production of marine fish remained relatively low. Although there

(a)　　　　　　　　　　　　　　　　　　(c)

(b)　　　　　　　　　　　　　　　　　　(d)

Figure 12.2　Australia; commercial marine culture: (a) Sydney rock oysters, Hawksbury River, New South Wales; (b) hatchery rearing Australian rock lobster; (c) harvesting barramundi; (d) Atlantic salmon grown in cages on farms.

had been significant investments in research and development for the culture of mullet, sea bass, bream, grouper, and many marine flatfishes, progress had been slow and success had not yet contributed significantly to the increased global total.

In addition to the rapid rise of the culture of the Atlantic salmon, the other most remarkable achievement of the 1980s was the increase in production of marine shrimps to meet continuous and increasing demands for crustaceans as a whole. About seventeen of these high-value species were under culture, the most important of which were the giant tiger prawn and fleshy prawn. Major industries were economically important in many Asian countries, and in Central and South America.

As aquaculture products, the freshwater crustaceans had not benefited by the popularity of marine shrimps. In particular, the high-value, giant river prawn had not realized its potential in aquaculture, due mostly to the low profitability of its culture. For investors, marine shrimp farming had a far greater return on investment. A shining exception was the freshwater lobster, or crayfish, which rose from nowhere to become a popular local delicacy wherever it could be farmed and a highly profitable farm-gate sideline for rice farmers or for farmers who had suitable wetlands.

The medium-high-value marine mollusks continued to be significant contributors to global aquaculture production. Both oyster and mussel production capitalized on traditional culture practices in European and Mediterranean countries, as well as in North America, and each contributed a significant part of the total. Mollusk production was also enhanced by the newer and rapidly expanding culture fisheries of clams—particularly the Japanese or Manila clam—and blood cockle. Although there was still some confusion about whether these new fisheries were genuinely "aquaculture," management practices and seeding of beds with spat reared in hatcheries had been increasingly evident in the last ten years in Asia.

The farming of marine algae in Asian countries, although on average low in value, continued to increase and make a major contribution to global aquaculture production. The brown seaweeds, which had been cultured traditionally for centuries, accounted for about two-thirds of the algae farmed annually, followed by the red seaweeds, and the more recently exploited green seaweeds. Top quality edible seaweeds, after processing, were very high in value, but the market was small and mostly confined to a few Asian countries. The greater part of the algal harvest was for industrial use in the food industry, and the individual value was low.

Despite the increase in market demand for new species in the last two decades of the millennium and the increased success in their culture, the traditional species still remained the most important in terms of production volume. Thus aquaculture, like agriculture, relied on a small number of crops for the greater part of its total production. In particular, production was dominated by freshwater fish, and from within that category, the various species of cyprinids probably accounted for half of the total aquatic animal production. That group was followed at some considerable distance by the salmonids, and even further behind by the tilapiines. The most important family of crustaceans was the penaeids, and production focused on four or five main species. The mytilids and ostreids were the principal mollusks.

Notwithstanding the large number of aquatic species under culture, the implication of the narrow reliance on a few key crops presented—continues to present—some genetic issues for the long-term future of the industry. Among these are the need to maintain genetic diversity of major food crops and to conserve genetic resources. Moreover, and most importantly, the restricted focus causes other needs to be overlooked or fails to exploit potentials of other species.

12.5   The reality of market forces

The successful development of the marine shrimp farming industry in Asia was one of the highlights of the 1980s. Production grew from a reasonable fifty thousand tonnes in 1980 to over 600 thousand tonnes by the end of the decade. From the start, the principal production was in Indonesia, the Philippines, and Thailand, but this was soon supplemented by growing yields from India and China. In Latin America, the star was Ecuador, although Mexico, and several of the smaller countries in Central America, contributed to the growing total.

The constant high prices for marine shrimp maintained the investment momentum, which was aided in Asia by several country loans provided by the World Bank and Asian Development Bank to their members. With the favorable returns on investment to the producers in Indonesia, and with the repayment of loans, the Asian Development Bank and the World Bank increased their lending to the aquaculture sector throughout the region. The timing was excellent. World harvests of shrimp had been gradually leveling off through the 1970s, and the increasing demand was filled by the new farmed resources.

The bubble burst at the end of the decade, as more producing countries (particularly India and China) came into operation, and the three principal markets, the United States, the European Union bloc, and to a lesser extent, Japan, went into a domestic recession. Prices fell, and many of the newer producers went bankrupt, because they were still repaying capital debts and had high production costs.

A similar scenario was being acted out in the salmon industry. Global harvests of the prime Pacific salmon, chinook and coho, were generally in decline. Their share of the existing and new markets was being exploited by the increasing availability of farmed Atlantic salmon from Norway, Scotland, Ireland, and Chile, along with farmed Pacific salmon from Canada, Japan, and the United States. At the start of the decade, farmed production was a modest 7 thousand tonnes, reaching a peak ten years later of just over 320 thousand tonnes. Again, prices were sustained longer than anticipated, because Norwegian producers began to freeze quantities of fish to protect the domestic industry. However, not all of the producers followed their example, and the excess supply led to a worldwide collapse of prices. From the boom of over $12 per kilogram, the price rapidly spiraled downward to less than $6 per kilogram. This bankrupted many of the newer producers who had just entered the industry, as well as some of the older ones who had recently invested in expansion.

However, the market situation at the end of the 1980s was not bad for everybody. In the United States, the catfish industry strengthened modestly but steadily, together with a new industry for largemouth bass. In Asia, there was phenomenal growth in production of the Chinese carps, led principally by expansion of production areas in China, together with a growing industry for the high-priced groupers. Increased production of mollusks, such as mussels and clams, was very much evident by the end of the decade.

The increase in market demand for many fish and fisheries products, as well as for more variety of seafood in the early, more affluent part of the 1980s resulted in experimental production of many lesser-known species, some of which became profitable. Other commercial interests have added to the list, for example, the culture of freshwater pearls for jewelry, certain mollusks for medical research, and crustaceans for medicinal extracts. In some African countries, cultured fish have been important to produce "relishes" that make leafy energy-foods more palatable. As a result, the last decade of the millennium saw a large number of new species brought under commercial culture, using similar or common techniques, all of which were in demand in the national or international markets.

Bibliography

Fishery Information, Data, and Statistics Service (FIDI) (1986) Catches and landings, 1984. *FAO Yearbook of Fishery Statistics*, Vol. 58. Food and Agriculture Organization of the United Nations, Rome.

Chapter 13

Modern Times (twenty-first century)

Abstract

The changing picture of aquaculture development around the world over the next hundred years is not easy to envision. At the 1883 London Great Exhibition, the raison d'étre *for developing fish culture was the poor state of marine fisheries, which is not comparable to the situation a century later. After five hundred years of peaks and troughs in public interest and support, aquaculture is not going to disappear; the general public wants its products. Despite anti-farming attitudes, the twenty-first century will see global self-sufficiency in food production becoming an international priority, but even population growth held to nine billion people still means that future quality of life is unpredictable. Fundamental constraints that limit what all countries will achieve in aquaculture over the next century create a race against the clock. It is vital that commitments to development are made now, before time runs out and opportunities are lost.*

13.1 Introduction

The changing picture of aquaculture development around the world from the present day and through the next hundred years is not easy to envision. Certainly, the progress of the twentieth century would have fulfilled many of the ambitions of the fisheries scientists and fish culturists who gathered at the Great Exhibition in London in 1883. However, the achievements by the century's end were certainly not fulfilled in the manner in which they planned or within the timeframe that they might have anticipated. Many of them would have been acutely disappointed to know that essentially little of technological significance

The History of Aquaculture. By C. E. Nash. Published 2011 by Blackwell Publishing Ltd.

in fish culture actually occurred for another seventy years after the exhibition. Moreover, they would be horrified to learn that the state of marine fisheries at that time, which was so much the *raison d'étre* for developing fish culture, was not so bad compared with the situation a century later.

Anticipating development in any technological field is necessary, whether or not in the end the forecasts prove to be right. Visualizing the future is an important part of planning; current trends are used by the professionals to set short-term and long-term goals for development, and to choose the right strategies to achieve them. Predicting the future is also stimulating for discussion. Consequently, it is always necessary to keep looking ahead, even though many of the predictions could be thought to be naïve by future generations.

13.2   An industry here to stay

The safest prediction is that aquaculture is not going to disappear, because a large part of the general public wants its products. This is no small feat, because for five hundred years, there have been great peaks and troughs in public interest and support. Now, at the start of a new millennium, there are indications that the foundation of a global industry is very stable. First, the consumers are responding favorably to the many benefits offered by farmed fish and shellfish. Speaking through their pocketbooks, consumers say that farmed commodities give value for money: that is, they offer a wide choice of very different products, and the products have consistently high quality. Consumers also appreciate the healthful image of seafood. Second, because of the quantities available and the almost year-round reliability of supplies, the processors, distributors, and marketers are now fully integrated into the industry. This is one of the most difficult but important barriers for any emerging food industry to overcome, and it took aquaculture over two decades to achieve. Finally, production is safely underpinned by a number of complementary technologies, all of which have been developed through competent research. Producers are raising crops competently, year after year, using a variety of production systems and practices that operate economically at an industrial level. As a result, the number of producers in many countries is far above the critical mass needed to create an economically important subsector that is of value to any government setting goals or making policies for agriculture or fisheries.

Another good reason for aquaculture's persistence in the twenty-first century is that the world needs it. Already, about one billion people depend on fish to meet their basic animal protein requirement. Aquaculture is an industry that produces food *directly* for human consumption, and particularly, food high in necessary and readily usable protein. Moreover, predictions for a strong and stable future based on human consumption of fish and shellfish products are not necessarily dependent on increased *per capita* consumption. On the contrary, it is possible that the annual consumption of fish and shellfish per capita could remain fairly close to the current 14.5 kilograms (live weight equivalent), even as the global demand for protein increases. On a global basis, that figure could even decline, if the combination of aquaculture production and capture fisheries

(a)

(b)

(c)

Figure 13.1 United States; marine sea cage, OceanSpar Technologies, LLC: (a) cage prior to submerging; (b) inside sea cage, gravity ring to hold net open; (c) outside sea cage, diver cleaning net.

harvests is not able to keep pace with global population growth. The United Nations currently predicts that nine billion people will require feeding by the year 2050. Using an average figure of only 15 grams of animal protein per person per day, which is far below that required by an active human body, this translates to a demand for 135 thousand tonnes of animal protein per day. That means, it would require about the equivalent of the current total annual aquaculture production of the United Kingdom to feed the world's population for just one day in the year 2050.

Aquaculture is one of the few options for increasing global production of fisheries in the twenty-first century. Fish and shellfish, produced either through aquaculture or by harvest of the natural resources, currently provide between 15% and 20% of all animal proteins for human consumption, and they will

Figure 13.2 Globally, the future of aquaculture lies in continuing innovation and offshore expansion. (Courtesy of Ocean Farm Technologies, Inc., Searsmont, Maine.)

continue to be important contributors to the enormous global demand. However, after quickly rising from 20 to 100 million tonnes since the ending of the Second World War, the annual harvest of the world's oceans began to level off about a decade ago. It has remained more or less static ever since. Therefore, it is up to aquaculture to produce any future increase in the global total.

Yet, that is not the whole story of what gives aquaculture an edge over harvesting the natural resources in the future. One principal benefit of farming fish and shellfish is that the majority of its products are sold fresh to the consumers. At the present time, only about forty million tonnes of the global harvest of fish and shellfish are sold in fresh form for human consumption, and half of this is derived from aquaculture. Almost as much again is frozen and stored for long periods, processes that can cause a loss of some of the seafood's nutritional value. The remaining part of the global catch is not for human consumption, but rather for reduction to fish meal and fish oils for use, among other things, in manufactured feeds for all farm animals, including fish. Although it is safe to predict that much more efficient use of the oceanic catch will be made in the present century, particularly by using processing wastes and bycatch, most of such efficiency will still not make more fresh products available to consumers. Therefore, perhaps the only real opportunity for an increase in fresh fish and shellfish production for human consumption is through increased growth and greater variety of aquaculture products. These can be produced by the current array of aquaculture technologies, which would be divided fairly evenly between aquatic farming practices and the enhancement of natural stocks.

Nonetheless, in spite of the stability of global aquaculture and the favorable portents for its future in the twenty-first century, there are some current trends that give cause for concern. There are growing divisions among the general public to segments that indicate by their everyday, direct and indirect actions that they have other priorities for the same resources, and that some among them do not want any more aquaculture, after all.

13.3   A rising storm of public opinion

Notwithstanding that farmers are now skilled in the management and operation of many different production technologies for aquaculture, sooner rather than later, there may be nowhere remaining for their farms. The sites most suitable for aquaculture, which is more environmentally compatible than many other development activities, are rapidly disappearing. Everywhere in the world, there are now growing numbers of different human interventions that are continuously increasing the competition for land and water, with little regard for the long-term consequences.

One social issue that is seriously beginning to constrain aquaculture and all other traditional rural industries in the twenty-first century is the clamor for an improved and less stressful quality of life. This human and quite reasonable goal is based on the desire for a family home, a well-paid job, and regular hours of work. Paradoxically, even these three simple ideals mean different things to different groups of people. In the developed world, improvement in the quality of life is invariably idealized in the minds of the public as working far away from the pressures of the cities and the pollution of heavy industries, perhaps to live and enjoy the peace of the rural environment. In the developing world, it is the opposite. Improvement is idealized as living in a city and enjoying its glamour, far away from the endlessly dull life in the countryside, and the drudgery of the family farm.

Another issue of growing concern is the preference for foods that are raised in very specific ways. In particular, this includes foods that are grown without the use of "chemicals"—that is, farm products that were not subjected to use of herbicides, pesticides, growth hormones, feed stimulants, or anything that might be considered unnatural. For such organically raised products, many are prepared to pay more. Some are also avoiding farmed products that they believe are reared inhumanely, such as chickens raised in batteries, veal from calves raised in small darkened rooms, and even fish grown in floating cages. Other groups of consumers are avoiding products that have undergone some genetic modification. The denominator in all these reactions that is becoming increasingly more common is a growing trend against intensive farming. For the world's expanding population, this is a serious trend. The majority of people in the world depend on intensive farming systems for food, as has been the case in the past, as well. Certainly, without the combined skills of the scientists, agricultural engineers, and farmers who intensified the production of cereals and many other crops, there would have been an even greater number of people who were not fed at all during the last forty years.

A third issue common to both developed and developing countries is the growing political pressure imposed on elected officials to devote more of the countryside and coastal regions to nature conservation, before available areas are overrun. Much of this has been a consequence of the economic growth in the tourism and recreation industries in recent years. A larger portion of the general public now has more free time. Human demographics show that life expectancy is increasing, and especially in the countries of the developed world, the work force is retiring earlier with money to spend on vacations and new homes along the coastal sun-belts. Ironically, the increasing damage to the environment caused by tourism has yet to become a political issue.

The consequences of migration are already apparent. Governments of developing countries that are facing large population growth and movement from the country are greatly overstretched to provide housing and jobs for their people. They are being compelled to take more and more agricultural land for housing and factories, and to clear forest lands to provide building materials and fuel at an even greater rate than ever before. These strategies are putting their traditional food-producing capacity at risk. Many have agricultural lands and water resources that are already underproductive. Invariably, the holdings and family farms are greatly fragmented and small, and irrigation waters are even now in short supply.

In developed regions, the situation is not as acute. Many countries have already made significant advances in population control. Furthermore, they have far greater resources of agricultural land per capita for the production of food, and they have managed forestry land for the collection and storage of water. They also have adequate technical and financial resources to conserve and reuse water. Nonetheless, although the issues are not as serious, competition among the various sectors for the available resources is equally as fierce. And public opinion counts with politicians. Quite illogically, the food production sectors are beginning to lose.

13.4   Defiance and indifference, and the changing face of agriculture and fisheries

Because of governmental support to restructure rural societies according to their individual population migration patterns, the nation's food producers, in any shape or form, are no longer a constituency that the politicians are quick and proud to represent. Once almost supreme in the ministerial portfolios of most governments, the national sectors of agriculture and fisheries are becoming more and more minimalized. Aquaculture, which may be a recognized subsector of either one, is damned by association.

In the highly-political and social milieu of competition for resources, the agriculture and fisheries sectors are losing. The use of water for aquaculture production, and particularly the necessary good quality fresh water that is fundamental for most practices, will be an increasingly low priority—much lower, for example, than water for drinking, industrial use, urban use, and crop irrigation. Furthermore, the quality of fresh water available for aquaculture will also be affected by loss of land-cover and by pollution. The face of agriculture and fisheries is changing not only through fair competition, but also through the general indifference and defiance of the public at large.

Despite the realistic and significant opportunity for the future growth of aquaculture in many countries, there is the strong possibility that its potential will not be fulfilled. This is because there is a growing indifference by the general public to aquaculture. Some people just do not want aquaculture—or for that matter, many other things. The reason is simple. The social code of "live and let live" has turned to dust.

Selfishness and greed are two unpleasant attributes that increasingly characterize the societies of the twenty-first century. Moreover, they are not necessarily associated only with the new social structures of the developed world. They are equally rife in developing countries. Selfishness is widely manifest by the growing and effective defiance by individuals to anything that runs counter to their own ideas and preferences. For example, there is the increasing perception by some members of the public that all intensive farming (and the keeping of animals in zoos or a traveling circus) is unkind to animals. Not content with simply exercising their democratic right to express their opinion by boycotting farmed products in the marketplace or supporting ecolabeling, more and more sympathizers are animating their views by vandalizing farms and releasing the captive livestock. Their actions are taken without regard for any consequences to the environment that they purportedly wish to protect.

Another growing body of the public is determined to defy any interference in their enjoyment of their personal lifestyles and pastimes. Some upbraid, for example, anything that constrains their supposed right to roam on any open land on foot or on wheels, or to navigate on any open water with their motorized pleasure craft; they object strongly to fences or farm structures that obstruct their free movement. Others admonish anything that might lower the value of their ever-encroaching suburban properties on rural lands. No matter who was there first, they object in particular to the smells and noises of farming practices and consider their farm structures unsightly. However selfish these public intrusions and accusations might be on the democratic rights of the farmers and landowners, or how unlawful the physical acts of defiance by extremists, through their numbers alone, these forces of public opinion are more than a match for the strength of any rural community—and particularly for one without any constituency in government.

Sociologists have observed that the beneficence of people in the postwar decades has been changing with each new generation. A social behavior based on the interest of the community, which was at its height in the 1940s and 1950s, has slowly been replaced with individual self-serving choice. This obvious transformation has been accelerated in no small way by many governments themselves. The once pragmatic and clear policies of postwar governments have become mired in topical but meaningless jargon. Ministerial statements are couched in a series of topical "buzz-words"—that is, slang or slogans in popular, though temporary use—such as *sustainability, integrated management,* and *stakeholders*. Perhaps worst of all has been the *precautionary principle*—a foggy resolve that enables any administration to abrogate responsibility for making any decision. Consequently, the future development of aquaculture, like that of many other new industries, is more and more constrained by the lack of government advocacy. As a result, it is being increasingly overwhelmed by negative public opinion.

In spite of the tremendous benefits of the so-called "super-highway" for making information available to everyone, public opinion is not always well informed. The information put into the public worldwide computer network system requires no authentication and guarantees no accuracy. Consequently, the super-highway has become a conduit for misinformation, deliberate or not, as well as for the facts that are also available. The computerized networks can be used to spread information without substantiation, and it is being taken at face value by gullible readers. Computer webs are becoming repositories for a plethora of *causes célèbres*; among these are several that target particular aquaculture systems and practices. Some of the most biased accusations, for example, have been unsubstantiated statements on the potential for farm fish in net-pens to escape and interfere with the general biodiversity of the local fauna, the excessive use of therapeutics in feeds that go into the environment, and the wastage of coastal mangrove forests for shrimp ponds.

Although some of these accusations have been justifiable areas of concern in the past, many governments have worked with their respective industries to minimize the problems, and significant progress has been made. Furthermore, the industry will always continue to resolve such issues, because they invariably have a bearing on the economic performance of the business, and it is therefore in its interest to find a solution. For example, new net-pen technologies and increased knowledge about sites have greatly reduced the number of escapes; the use of therapeutics has been almost completely replaced by the development and use of vaccines; and for the most part, regulatory authorities are confining shrimp production to specific coastal zones without endangering coastal mangrove forests.

Despite the efforts made by the aquaculture industry to improve its global image, the growing indifference of the public will continue to constrain growth. However, aquaculture is not fighting this issue alone. The public as a whole does not seem to care anymore about the origins of the food they eat, who produces it, or for the land or water on which it is grown. The fruits of the Green Revolution, developed by teams of scientists, engineers, and farmers, are becoming controlled by the cold, disinterested hands of the accountants of the large grocery chains that now monopolize about 80% of the consumer's food basket. To this type of accountant, those called "bean counters," one supplier is as good as another, providing the price is right. Consequently, farmers are now expected to grow and sell their products, whatever they are, at little more than cost and frequently less, whereas consumers are expected to pay anything between four and ten times that price for transportation and sale at a retail market, or possibly between twenty and fifty times that price if the product is processed in some way to add value. And the consumers do pay, with no thought for the producer who has nurtured and sweated over his crop and who has continued to fulfill the role of steward of the countryside in what little spare time was left to him in a seven-day week.

The truth is that farming today, in almost any form, is an unrewarding and thankless task. The younger generations have been realizing this for some time, and they have been selling the land and getting out of the family farm. In some ways, the traditional farming families share some of the blame for the depressed state of the business. They appear to enjoy a comfortable lifestyle, but continuously complain about low prices, the need for financial subsidies, and of course the problems of the weather. Yet, most of the blame has to rest with government administrations, particularly those working within a regional economic group such as the European Union, which requires continuous fiddling and tinkering within the system to rationalize equality among the members. Consequently, there have been policies supporting incomprehensible grants and subsidies, including subsidies for not growing a crop and even for taking land out of production. There have been schemes to control supplies, with stockpiles of products the size of mountains and lakes. Finally, there have been schemes to turn over productive land and coastal areas to developers, possibly to be lost forever.

The same scenes have been played out in the fishing industries around the world. The governments have dickered with fisheries management policies, catch quotas, licenses, and vessel buy-back schemes to take fishermen out of the industry. This has left many national fishing fleets in ruin and engendered strong hostility by displaced fishermen against the new generation of fish farmers, whom they accuse of taking away their livelihoods.

Although governments' concerns for overproduction of food and overharvesting of fish may have been necessary, and their policies well meaning, the results have had a drastic domino-effect on the economic life in rural and coastal villages. Farming or fishing, and often both, were the principal economic bases of such remote communities, and they have slowly disappeared. Moreover, their disappearance is irrecoverable. The prime land has been developed, and the tythed cottages sold and modernized by the weekenders or holiday-makers who contribute little or nothing to the tax base or to the village life.

13.5   How will it all end?

In spite of the current antifarming attitude by society at large, it should be safe to predict that sometime during the twenty-first century, global self-sufficiency in food production will once again become an international priority. If population growth in the near future could be curtailed to some degree, and the world population held to the projected figure of nine billion people, the future quality of life in many countries will still be unpredictable. In the last decade alone, the increasing frequency of major incidents of famine and disease has shown how entire regions are becoming increasingly vulnerable to any number of natural or manmade disasters. Not surprisingly, the frequency and consequences are greater in the presently developing countries of Asia, Africa, and Latin America than in the developed countries of Greater Europe and North America. This is because the populations of these three continents are already large, and their growth rates are far from being under control. The United Nations predicts that almost the entire addition of three to four billion more people in the world in the next fifty years will be in the developing countries on these continents. The consequence is continuing urbanization, whether a result of economic growth in the developed countries or poverty in the developing ones. Over half the world's population lives in urban areas, and that percentage will continue to increase.

Despite the heavy demands on all resources and the conflicting competition for their use, most countries will endeavor to maintain a production base for some degree of self-sufficiency in food. Global food production will also be augmented to some extent by new high-yielding varieties of crops through advances in genetic manipulation, new foods from yeast and algae, and some synthetic food products. However, growing and distributing food to the regions rapidly expanding their populations will mean that many countries in Asia, Africa, and Latin America will remain on a collision course with nature.

With due deference to the principles of the Reverend Malthus, the future for aquaculture is far from bleak. In the highly populated and poorer regions of the world, where markets will demand large quantities of cheaper fish to provide animal protein, freshwater farming will continue to be the priority. This is because the freshwater environment is more easily managed, and there are various key fish species that will become truly domesticated, capable of growth to pan size in a short period, and not entirely dependent on artificial feeds. They are also familiar to most consumers, and inexpensive. However, freshwater farms will be designed around the conservation of water and its multiple use. Farm production will be increased by building on the practices now greatly evident in China, specifically integrating fish husbandry in ponds with animal husbandry, side by side with production of vegetables and crops, or by increasing pond productivity with the addition of processed household and industrial wastes. Some of these freshwater farms will be allocated water resources that have not yet been used, and others will be allocated water that has already been used several times before, including sewage waters. However, farms will have to adhere to practices that greatly restrict water loss, and pass on their own nutrient-enriched water for other uses, particularly for irrigation of crops and vegetables.

In regions where there are markets for high-quality freshwater fish, production will become increasingly dependent on land-based farms with totally enclosed units and capability to recycle fresh water after cleansing. Of necessity, to remain economical, these farms will be for high-value species because of the high energy costs of recycling and artificial feed. Many production systems based on recycling technologies have already been developed. Now they are simply waiting for the time when the demand for fresh fish will make them totally economical to operate.

In the less-populated and richer regions, the markets will always remain strong for high-quality saltwater fish and shellfish. However, many countries might not be able to meet all the demand because of the diminishing number of sites. Coastal brackish-water areas, potentially the most suitable for farming due to their natural productivity, will continue to be irretrievably lost. Many shallow estuaries and low-lying coastal areas, which for centuries provided seafood for the early cultures, began to disappear in the eighteenth century. If not first drained and used as grazing land or for crop production, they were reclaimed for heavy industrial use, because they were flat and convenient to the port. Later, they were reclaimed for urban development around the industries, and more recently they have been lost to the demands of tourism and recreation. Many estuaries that remained unpolluted and unscathed by the end of the twentieth century have been wisely conserved as nature reserves and wildlife sanctuaries.

Special zones for aquatic farming will be much more difficult to obtain or protect in countries where the competition for coastal resources is fierce. In the poorer countries, where economic pressures are acute, there has been a continuous migration of people from the land to the towns to find work, and particularly to coastal towns, if there are growing tourist industries. There, the pressures of population and economic growth will mean that the most accessible and sheltered coastal zones ideal for farming will be lost through the pollution that emanates from increasing urbanization. In richer countries, where economic pressures are less serious and the migration to the coast has been mostly for lifestyle and recreation, any farming potential will be lost in the face of ribbon-development and uncontrolled coastal sprawl, and the mindset of the new settlers who are typically against anything that detracts from their environment, unless it is recreational.

Some countries do still have coastal areas that remain potentially available for further farming use. But in order to realize these remaining opportunities, there must be well-established market demand for farm products and the political and social will to use these areas. This is always easier to establish in countries that either abound in such coastal resources or that have coastal communities with depressed economies. Consequently, in the future, aquaculture development within the coastal zone will be predominantly integrated into a basket of economic activities developed by communities, rather than by individual entrepreneurs.

True marine areas, which are offshore over deeper waters, will not be affected, provided that aquatic farming development conforms to some old and some new regulations. Away from the sheltered shoreline, the offshore farms of the future will be made up of submerged rigid and semirigid structures, held in place by spar buoys and tensioned by anchors. They will be located in sites away from any shipping so that they do not become navigational hazards. There will, however, be occupational hazards, because daily operation and maintenance will make exceptional demands on the employees. Ideally, a number of such farms could be sited around an ocean rig of some type, which would act as a service platform. However, the investment cost will probably remain too prohibitive for individual investors, and these high-risk ventures will be the prerogative of the multinational corporations.

As a result, if the global markets for fish and shellfish products remain strong, but margins for the farmers remain thin as operational costs for higher-value products continue to rise, then countries will turn more and more to encouraging stock enhancement through the application of aquaculture technologies. In some respects, this will be turning back the clock a hundred years. This would no doubt please many of the early fisheries scientists who might be watching from afar and waiting for their ideas to be implemented at last. However, the enhancement schemes of the future will not be liberal releases, but rather will be controlled releases in areas that can be managed: that is, freshwater lakes and reservoirs, and coastal sites enriched by artificial reefs or offshore structures.

Fisheries enhancement is a practical solution to future humanitarian need for animal protein that will not please everyone. Many conservationists oppose all human intervention that might destroy the natural biodiversity of an ecosystem. Therefore, unless the fish and shellfish being released from hatcheries and rearing sanctuaries are almost identical to wild stock in genetic makeup, behave like wild fish, and are infertile, then there will be continued opposition to such schemes, however great the need for food. This would be unfortunate, because many systems are already proving to be successful. In Japan, for example, the fishermen, farmers, and scientists long ago started to work purposely together. As part of a national program for "selective adaptation of a fishermen's society," they developed ecologically compatible enhancement systems, closely integrated so that species derive mutual benefits. They also built fishing grounds around coral reefs; opened coastal recreational fishing centers; developed innumerable small-scale fisheries; and targeted remote communities for special help. The result is that over half the large harvest of farmed fish and shellfish in Japan is produced by such community-oriented cooperative activities. But all this is no surprise. Successive postwar Japanese governments always recognized that food production was the highest of priorities; therefore, the Japanese farmers and fishermen were given every possible support to make a useful living while contributing to the national cause.

In conclusion, it is clear there are several directions to choose for the development of global aquaculture in the next century. Selection, however, depends on the need and the public commitment. It is public demand, after all, that drives growth, not politicians or technologists, and the markets created by the various peoples of the world will continue to be very different. Therefore, some countries will make some production systems work better than others.

All countries have the same fundamental constraint that can limit what each will achieve in the next century: *time itself*. It is vital that many commitments to development are made now, before the time runs out and opportunities are lost. Aquaculture is in a race against the clock.

Appendix

Abbreviations

Table A.1 Abbreviations, Acronyms, and Initialisms

Item	Definition
ADCP	Aquaculture Development and Coordination Programme
AQUACOP	The aquaculture team of the Centre Océanologique du Pacifique
CARE	Cooperative for American Remittances to Europe (when founded in 1945; today, it stands for Cooperative for Assistance and Relief Everywhere, Inc.)
ECU	European currency units
EEC	European Economic Community
FAO	United Nations Food and Agriculture Organization
FFI	Fish Farming International
FIDI	Fishery Information, Data, and Statistics Service
ICLARM	International Centre for Living Aquatic Resources Management
NMFS	National Marine Fisheries Service
NOAA	National Oceanic and Atmospheric Administration
PACAQUA	Pacific Aquaculture Caucus
PNP	Private Non-Profit Aquaculture Associations
PVC	Polyvinylchloride
SEAFDEC	Southeast Asian Development Center
UNDP	United Nations Development Programme
VSO	Voluntary Service Overseas (United Kingdom)

Glossary

Table A.2 Glossary of Terms in Text

Term	Definition
Actinomycete	A group of gram-positive bacteria, which includes some of the most common soil, freshwater, and marine microorganisms that play an important role as decomposers of organic material, such as chitin and cellulose; currently renamed actinobacteria. (After Wikipedia.org)
Anadromous	Describes fish that ascend upriver from the sea to spawn.
Aquahusbandry	Selective breeding and raising of cultured aquatic species.
Artemia	Genus of brine shrimp.
Bakufu	A period of military rule in Japan by a hereditary shogun, as opposed to rule by the imperial court and the emperor, in this case prior to the end of the Edo Bakufu 1603–1868. (From Answers.com)
Beam trawl	From a fourteenth century fishing net design called a *wondyrchoum*, which used a three-meter wooden pole or "beam" to hold open the mouth of a net six meters long. Originally used by fishing vessels under sail, the beam trawl was more efficient and common in the nineteenth century with steam vessels, which could use steam power for winches to more easily raise and lower the heavy beams and had the power to drag the heavy beam and trawl over the sea bottom. (After Wikipedia.org)
Boucholeur (*bouchot* system) (French)	A system that originated around 1035 in the Bay of Aiguillon in France by Walton, the shipwrecked captain of an Irish barque. It uses rows of wooden poles set in the mud perpendicular to the shore at close intervals to support culture of blue mussels (*Mytilus* sp.). At first this was on woven branches of trees, but more recently on netting from various natural and manufactured materials. ([1892] *Proceedings and transactions of the Liverpool Biological Society* 7, 121.T. Dobb & Co., Liverpool, England)
Carposphere	Spore of red algae, diploid stage.
Chemical attractants (in fish feeds)	Feed additives to increase palatability to the fish in mouth-feel, taste, and smell.
Growth promoters (in fish feed)	Feed additives to increase the efficiency of uptake in the fish's gut through optimizing absorbance of nutrients, minerals, and vitamins.
Cichlids	Fish of a large and diverse family, Cichlidae, which includes some food fish such as tilapia, popular aquarium species such as angelfish, oscar, and discus, and some game fish.
Ciguatoxin	One of a group of toxins that bioaccumulates in the flesh of certain reef fishes in tropical and subtropical waters, and that is heat resistant—that is, the fish cannot be detoxified by conventional cooking.
Colossomids	Fish mostly of genus *Colossoma*, some species of which are captured, also cultured in South America and elsewhere as a popular food item and as ornamental fish. At least some are disk-shaped, laterally compressed.

Table A.2 Glossary of Terms in Text (*Continued*)

Term	Definition
Conchocelis phase	Originally thought to be a separate species and referred to as *Conchocelis rosea*, it is now known to be part of the reproductive diploid stage of the red alga (seaweed) *Porphyra* sp., which has three distinctive phases in its life cycle.
Cultchless seed	Single oysters produced in shellfish nurseries as seed from the hatchery's own oyster larvae, by allowing them to attach and metamorphose individually on very finely ground particles of oyster shell. The traditional method employs larger numbers of either hatchery or wild larvae, which are allowed to attach to larger pieces of oyster shell or other artificial substrate suspended in tanks or open waters.
Cyanobacteria	Phylum of bacteria that obtain their energy through photosynthesis, also called blue-green algae or blue-green bacteria.
Cyprinid	Fish in a family (Cyprinidae) that includes carps, minnows, and their relatives; mainly freshwater species.
Depuration	Process of treating shellfish by holding them in tanks of clean seawater under certain conditions to maximize the natural filtering activity, which expels the intestional contents and thereby rids them of any contaminants they might have gathered as they fed from the water column.
Dry method (of fertilization)	A method of induced spawning by which the sperm of the males of the species is poured onto the eggs without contact of water (water would inhibit fertilization). After fertilization, the eggs are hardened with appropriately timed exposure to water before incubation.
Elver	The stage of juvenile eel development after the "glass," or transparent stage, the stage of development where they migrate to the sea.
Eyed egg	Salmonid eggs at approximately thirty days after fertilization, when they develop dark spots (eyes).
Feddan (Egyptian)	An Egyptian unit of area equivalent to 1.038 acres (0.42 hectares), originally based on the amount of land a pair of oxen could farm, which varied from area to area. (After Wikipedia.org)
Fish seed	Hatchling, spawn, fry, or fingerling—that is, any of the life stages of young fish used to introduce a species to a location.
Franklin	Landowner who was not a noble.
Gadoids	Soft-finned fish of family Gadidae, including cod and hake.
Garum	A concentrated fish sauce, made through a fermentation process.
Geosmin	An organic compound with a distinct earthy flavor and aroma, responsible for the muddy smell in bottom-dwelling freshwater fish such as carp and catfish. Cyanobacteria produce geosmin, which concentrates in the skin and dark muscle tissue of fish that consume them. (After Wikipedia.org)

Table A.2 Glossary of Terms in Text (*Continued*)

Term	Definition
Gillies	Men and boys who waited on or attended to the needs of others in fishing or hunting activities on Scottish estates. Their roles might also have included that of river keeper and cultivator of trout and salmon stocks on portions of rivers and streams found on the estates.
Gourami	Members of a family of freshwater fish native to Asia. Some species are popular aquarium fish (e.g., Siamese fighting fish, kissing gourami), others are edible.
Great Cultures	Historical context: those cultures present prior to the fifteenth century.
Great Withdrawal	A change in policy in 1433 in Chinese history, Ming Dynasty, when Chinese scholar bureaucrats outlawed overseas travel and importation of foreign goods, arguing that the truly cultivated need no acclaim from barbarians to be certain of their own superiority. This cut short China's age of discovery. Between 1405 and 1433, seven expeditions by the emperor's ambassador, Cheng Howere, were made in splendid flotillas of several hundred junks, some immense, nine-masted ships, 444 feet long, which could safely venture the nearly 10,000 miles to Zanzibar. The voyages had been made to broadcast Ming Dynasty Chinese technical triumphs to all of the civilizations in the known world.
Head-line	In fishing gear, the rope or wire rope at the top of a fishing net or trawl. It may be composed of a nonbouyant wire rope or even a chain with additional units of buoyant material for positive floatation to keep it at or above the top edge of the netting.
Hibi (Japanese)	Primitive devices built of bamboo poles and brushwood for collecting and growing seaweed in seawater bays.
Hypophysation	A hormonal manipulation of fish using the pituitary gland (part of the brain and endocrine system) to allow breeding of some species that otherwise do not spawn in a certain season or at all in captivity for aquaculture.
Kuruma abi (Japanese)	A kind of shrimp. The words in Japanese mean the "shape of a wheel" from the way the exoskeleton of the shrimp is arched.
Lab-lab (Javanese)	Biological mat of microfauna that accumulates at the bottom of saltwater fish ponds, mangrove swamps.
Latin square	In statistical analysis, an $n \times n$ table filled with n different symbols in such a way that each symbol occurs exactly once in each row and exactly once in each column (from Wikipedia.org). In this case, a six-by-six geometric array of ponds to allow for up to six fully randomized treatments in fish-farming experimental designs, enabling both statistical and combinatorial analysis of the results.
Laver	Any of several common red algae (genus *Porphyra*) with edible fronds.
Loko kuapa (Hawaiian)	Constructed tidal ponds with permeable coral walls.
Loko 'umeiki (Hawaiian)	Constructed tidal ponds, with stone and coral walls built on reefs, with one-way seawater gates.

Table A.2 Glossary of Terms in Text (*Continued*)

Term	Definition
Menhir	A large, upright, prehistoric monumental stone, standing alone or with others, as in an alignment such as those seen chiefly in Cornwall and in Brittany. The world's largest alignment of over one thousand stones is found at Carnak in southern Brittany.
Mytilids	Members of the family Mytilidae composed of the small to large sea mussels of the world. Mytilids include the well-known edible sea mussels.
Nanoplankton	One size-range of planktonic organisms, consisting of two to twenty-micron creatures such as diatoms and flagellates.
Nauplius	The first larval stage of development of some crustaceans.
Nori (Japanese)	Edible seaweed species of the genus *Porphyra*.
Oracle bones	Pieces of bone or similar material used in divination, mainly in ancient China (Shang Dynasty).
Ostreids	Members of the family Ostreidae, which are the true oysters, and include all species that are commonly eaten under the name of oyster.
Otter trawl	A type of net on which the mouth is held open by two rectangular panels or "otter boards," each attached to one end of a towing bridle. When they are towed, the two otter boards are forced apart by water, holding the mouth of the net open without using a rigid cross member as in a beam trawl.
Parr	A stage in salmon development when the fish are a few months or over a year old and develop certain markings; they then feed and grow in fresh water shallows until they undergo smoltification.
Penaeids	Members of the family Penaeidae, which are economically important, widely farmed animals that are actually prawns, though commonly referred to as shrimp.
Pescheria (pl. -ie) (Italian)	Fish shop, fish market.
Peschiere (pl. -ae) (Latin)	Fishponds.
Piscina (pl. -ae) (Latin)	Simple fishpond.
Piscinarii (Latin)	Those fond of fishponds.
Polyculture	In aquaculture, the practice of culturing multiple organisms together in the same ponds, creating a system that can more efficiently use resources than can a single species, and that therefore maximizes production of the fish.
Proteolytic	Adjective describing that which causes splitting of proteins or peptides by the action of enzymes.
Salmonids	Fish of the family Salmonaediae, which arose from three lineages: whitefish (Coregoninae), graylings (Thymallinae), and the char, trout and salmons (Salmoninae).
Salt pan	Depression in the ground near the sea, or expanse of flat land usually in a desert, from which salt and other minerals can be harvested through evaporation of brines.

(Continued)

Table A.2 Glossary of Terms in Text (*Continued*)

Term	Definition
Set net fishery	Any fishery where gill nets are fixed to the bottom by anchors or to the land by lines while the nets are employed in fishing. The buoyant, horizontal head-line at the top of the net is supported by floats, and a weighted foot-line holds down the bottom horizontal edge of the net. Nets can be located in intertidal areas and emptied when the tide goes out while the net is lying on the beach. In shallow subtidal waters, the net is emptied by pulling it over one side of a boat, removing the fish and allowing the net to go back in the water over the other side of the boat. Set nets can also be deployed by suspending them off the bottom from anchors and returning them to the surface for emptying in the boat.
Shik (Chinese)	A traditional Chinese unit of market or Chi measure. Also Shi and Chi. Six chi equal one bushel or 35.24 liters, one Chi equals 5.87 liters.
Siwakan (Javanese)	Artificial freshwater fishpond.
Smolt	A young salmon at the stage when it becomes covered with silvery scales and first migrates from fresh water to the sea.
Stew pond	Handy live fish holding pond for the kitchen.
Tambak (Javanese)	Artificial saltwater fishpond.
Thallus (pl. thalli)	From Latinized Greek, the whole vegetative body of some organisms, such as fungus and algae; although not differentiated to organs, it can have functional structures that resemble the stem, leaves, or roots of vascular plants. In seaweeds, the thallus is sometimes called a frond.
Tilapiines	The Tilapiini is a tribe within the family Cichlidae commonly known as tilapiine cichlids. Most of the taxa herein are called "tilapias," a diverse and economically important group containing the genera *Oreochromis*, *Sarotherodon* and *Tilapia*. (From Wikipedia.org)
Tonne	Metric ton = 1,000 kilograms (2,200.62 pounds), distinguish from ton (short ton) in American usage = 2,000 pounds.
Triclinium (Latin)	A place where patricians lounged, talked, ate.
Tythed cottages	In British history, such cottages were bought with tythe (tithe) funds derived from agriculture by the community, and given for the poor to live in. The poor then contributed back to the community through agriculture or other activities. The cottage was not on the tax roles, because it was church or community property. Recently, many tythed cottages have been renovated and leased out or sold as country weekend vacation places for people from the city.
United Nations Codex Alimentarius	A collection of internationally recognized standards, codes of practice, guidelines and other recommendations relating to foods, food production and food safety. (Wikipedia.org)
Valleum (Latin)	Paling, a fence made of pointed stakes.
Valli (Latin)	Fenced enclosure(s) of lagoons, ponds.
Vivariae (Latin)	Ponds for holding living organisms.
Zoea	Free-swimming larval stage of crustaceans.

Species lists

Table A.3 lists by common and scientific name the species of fish, shellfish, crustaceans, and other miscellaneous organisms mentioned in the text. Because this book explores a history that spans millennia and because it is based on a variety of records and documents in which both folk taxonomy (common names in local vernacular) and an evolving scientific classification were used, the precise identity of some species to which the written resources refer cannot be determined. The author made the best possible estimate of the species that occurred at the particular periods and in the geographical locations described. There is a degree of uncertainty, because creatures do not always follow the established distributions and regional preferences described by scientists. Rather, they respond to constantly changing climate and ocean conditions.

For a more complete species reference, please see the United Nations Food and Agriculture Organization (FAO) website (www.fao.org/docrep/w2333e/ W2333E00.HTM), where the FAO Fisheries Circular No. 914 FIRI/C914, *List of Animals Species Used in Aquaculture*, can be found.

Table A.4 lists by common and scientific name the species of fish, shellfish, and other organisms portrayed in the first-century Pompei mosaic of rocky-coast marine fauna of the Mediterranean Sea (Figure 1.1), as identified in Annamaria Ciarallo's text, *Pompei e le acque* (2006), pages 30 and 31. Included with the sea life is a piscivorous bird, the Eurasian kingfisher, "perched as if ready to capture some fish" (Ciarallo, p. 29).

Table A.3 Species mentioned in the text.

Common name	Family, genus, species
Fish	
Aloes	*Clupea alosa*
Amberjack (jack)	*Seriola* sp.
Barb (Java barb)	*Barbonymus gonionotus*
Barbel	*Barbus barbus*
Barracuda	*Sphyraena* sp.
Bass	
Sea bass	Unknown
Smallmouth bass	*Micropterus dolomieu*
Striped bass	*Morone saxatilis*
European sea bass	*Dicentrachus labrax*
Bream	*Sparus aurata*
Gilthead bream (daurade)	
Buffalo fish	*Ictiobus* sp.
Carp	
Common carp	*Cyprinus carpio*
Silver carp	*Hypophthalmichthys molitrix*
Mud carp	
Grass carp	*Cirrhinus chinensis*
Bighead carp	*Ctenopharyngodon idella*
Wild goldfish	*Hypophthalmichthys nobilis*
Chinese carps	*Carassius auratus*
Indian carps	Various**

(Continued)

Table A.3 Species mentioned in the text. (*Continued*)

Common name	Family, genus, species
Great carps	Various∗∗
Golden carps	*Cypinus carpio* (possibly)
Catfish	
Sheat-fish	*Silurus glanis*
Silure (wels)	Silurus glanis
Pangasius	Pangasius *spp.*
Channel catfish	Ictalurus punctatus
Blue catfish	Ictalurus fucatus
Char (*ombre chevalier*)	*Salvelinus alpinus alpinus*
Arctic char	*Salvelinus alpinus*
Cod	
Atlantic cod	*Gadus morhua*
Dace	*Leuciscus* sp.
Eel	Family Congridae (>100 species)
Conger eel	*Conger* sp.
River eel	*Anguilla japonica*
Eel-pout	Unknown (possibly *Lota lota*)
Fera (lake herring)	*Coregonus fera*
Flatfish	Family Pleuronectidae (see halibut, sole, turbot, plaice, flounder) (101 species)
Flounder	Family Paralichthyidae
Gourami (Siamese gourami)	*Trochogaster pectoralis*
Grayling	*Thymallus thymallus*
Grouper	Family Serranidae Subfamily Epinephelinae (among others)
Haddock	*Melanogramus aeglefinus*
Halibut (Atlantic)	*Hippoglossus hippoglosus*
Herring	*Clupea harengus*
Lamprey	*Lampetra* (freshwater)
Mackerel	*Scomber* sp.
Milkfish	*Chanos chanos*
Minnow	*Phoxinus phoxinus*
Mullet	
Red mullet	Family Mugilidae
Gray mullet	Family Mullus
Parrot wrasse	*Scarus cretensis*
Perch	*Perca*
River perch	*P. fluviatilis* (probably)
Pike	*Esox* sp.
Pike-perch	*Sander* sp.
Pilchard	*Sardina pilchardus*
Piranha	Family Characidae, several genera
Plaice	*Pleuronnectes platessa*
Pollock	*Pollachus chalcogrammus*
Pompano	*Trachinotus* sp.
Puffer fish	Family Tetraodontidae
Roach	*Rutilus rutilus*
Rabbitfish	*Siganus* sp.
Salmon	
Atlantic salmon	*Salmo salar*
Danube salmon	*Hucho hucho*
Pacific salmon	*Onchorhynchus* sp.
Chum salmon	*Oncorhynchus keta*
Chinook salmon	*Oncorhynchus tshawytscha*
Coho salmon	*Oncorhynchus kisutch*

Table A.3 Species mentioned in the text. (*Continued*)

Common name	Family, genus, species
Cherry salmon	*Oncorhynchus masu*
Pink salmon	*Oncorhynchus gorbuscha*
Shad	*Alossa sapidissima*
Sole (Dover sole)	*Solea solea*
Starlet	*Acipenser ruthenus*
Sturgeon	*Acipenser* sp.
Swordfish	*Xiphias gladius*
Tench	*Tinca tinca*
Ten-pounder	*Elops sarus*
Tilapia	
Red tilapia	*Orechromis* sp.(probably)
Nile tilapia	*Oreochromis niloticus*
Trigger fish	Family Balistidae
Trout	
Rainbow trout	*Oncorhynchus mykiss*
Brook trout	*Salvelinus fontinalis*
Brown trout	*Salmo trutta*
European trout	*Salmo trutta* (possibly)
Steelhead trout	*Oncorhynchus mykiss*
Tuna	Family Scombridae, mostly *Thunnus*
Turbot	*Psetta maxima*
Whitefish	*Coregonus lavaretus*
Yellowtail	*Seriola quinqueradiata*
Shellfish	
Abalone	*Haliotus* spp.
Clam	
Japanese clam	Unknown
Manila clam	*Venerupis philippinarum*
Cockle	
Blood cockle	*Andara granosa*
Limpet, parasitic slipper	Crepidula fornicata
Mussel	
(Chinese production)	*Mytilus galloprovincialis*
Green mussel	*Perna viridis*
Thick shell mussel	*Mytilus coruscus*
Oyster	
European flat oyster	*Ostrea edulis*
American cupped oyster	*Crassostrea virginica*
Pearl oyster	*Pinctada* sp.
Scallop	Family Pictinidae
Tingle, American whelk	*Urosalpinx cinerea*
Crustaceans	
Brine shrimp	*Artemia* sp.
Crawfish	*Procambarus clarkia*
Crab, softshell Dungeness	*Cancer magister*
Shrimp	
Marine shrimp	Penaeus *sp.*
Giant freshwater prawn	Macrobrachium rosenbergii
Kuruma abi marine shrimp	Marsupenaeus japonicus
Giant tiger shrimp	Penaeus monodon

(*Continued*)

Table A.3 Species mentioned in the text. (*Continued*)

Common name	Family, genus, species
Brown shrimp	Penaeus aztecus
White shrimp	*Penaeus setiferus* (probably)
Pink shrimp	*Penaeus duorarum*
Fleshy prawn	*Fenneropenaeus chinensis*
Seaweeds	
Laver	*Porphyra laciniata, P. umbilicalis*
Nori **(Japanese)**	*Porphyra yezoensis, P. tenera*
Miscellaneous	
Frogs	Families of order Anura
Sea squirts	Family Ascidiacea
Sponges	Animals of the phylum Porifera
Turtles	Reptiles of the order Testudines

*Common, bighead, silver, and grass carps are Chinese carps.
**Catla (*Catla catla*), rohu (*Labeo rohita*), mrigal (*Cirrhinu mrigala*), kalbasu (*Labeo calbasu*) are Indian major (great) carps.

Table A.4 Species portrayed in Pompei mosaic (Figure 1.1).[a]

Category	Common name	Family, genus, species
Fish	Common torpedo (electric ray)	*Torpedo torpedo*
	Small-spotted cat shark	*Scyliorhinus canicula*
	Nursehound	*Scyliorhinus stellaris*
	Mediterranean moray eel	*Muraena helea*
	Striped red mullet	*Mullus sermuletus*
	Gray mullet	*Mugil auratus*
	Gilthead bream	*Sparus aurata*
	White sea bream	*Diplodopus sargus*
	Bogue	*Boops boops*
	Dusky grouper	*Epinephelus guaza*
	European sea bass	*Dicentrachus labrax*
	Painted comber	*Serranus scriba*
	Gurnard	*Trigla* sp.
	Red gurnard	*Aspitrigla cuculus*
	Black scorpionfish	*Scorpaena porcus*
Crustaceans	Caramote prawn	*Peneatus kerathus*[b]
	Rock lobster (sea crayfish)	*Palinurus vulgaris*
	Barnacle	*Balanus* sp.
	Conch	*Strombus* sp.
Mollusks: Gastropods	Purple-dye murex	*Murex brandaris*
Cephalopods	Common octopus	*Octopus vulgaris*
	European squid	*Loligo vulgaris*
Birds	Eurasian kingfisher	*Alcedo atthis*

[a]Identification of species from the following publication: A. Ciarallo (2006) *Pompei e le acque: il fiume e il mare*, Ministero per i Beni e le Attivita Culturali, Soprintendenza Archeologica di Pompei, Electa Napoli s.p.a., Naples, Italy, pp. 30–31.
[b]*Peneatus kerathus* may be elsewhere known as *Penaeus kaerathus*.

Suggested Reading

Resources

Aflalo, F.G. (1904) *The Sea-Fishing Industry of England and Wales*. Edward Stanford, London.

Alward, G.L. (1932) *The Sea Fisheries of Great Britain and Ireland*. Albert Gait, Grimsby.

Association of Scottish District Salmon Fishery Boards. (1977) *Salmon Fisheries of Scotland*. Fishing News Books Ltd., Farnham.

Bailey, R.S. and Parrish, B.B. (1987) *Developments in Fisheries Research in Scotland*. Fishing News Books Ltd., Farnham.

Blegvad, H. (1932) Plaice transplantations. *Journal du Conseil* 8(2), 161.

Buckland, F.T. (1863) *Fish Hatching*. Tinsley Brothers, London.

Chen, T.P. (1976) *Aquaculture Practices in Taiwan*. Fishing News Books, London.

China Fisheries Society (History Research Association). (1997) *Fan Li on Pisciculture*. The Agriculture Publishing House, Beijing.

Clements, J. (1988) *Salmon at the Antipodes*. John Clements, Australia.

Cutting, C.L. (1956) *Fish Saving*. Philosophical Library, New York.

Davy, F.B. (1991) *Mariculture Research in Japan: An Evolutionary Review*. International Development Research Center, Ottawa.

De Landgraf, J. (1916) *Fishing and Fish Culture in Hungary*. International Institute of Agriculture, Printing Office of the Institute, Rome.

Dobrai, L. and Pékh, G. (1989) *Fisheries in Hungary*. Ministry for Agriculture and Food, Budapest.

Dunfield, R.W. (1985) *The Atlantic Salmon in the History of North America*. Canadian Special Publications in Fisheries and Aquatic Sciences. Department of Fisheries and Oceans, Ottawa.

Dyk, V. and Berka, R. (1988) Major stages of development in Bohemian fishpond management. *Práce Vúrh Vodňany* 17, 3–44.

Erickson, C.L. (2000) An artificial landscape-scale fishery in the Bolivian Amazon. *Nature* 408, 190–193.

Graham, M. (1948) *Rational Fishing of the Cod of the North Sea. The Buckland Lectures for 1939*. Edward Arnold & Co., London.

Graham, M. (1956) *Sea Fisheries: Their Investigation in the United Kingdom*. Edward Arnold (Publishers) Ltd., London.

Gross, F., Orr, A.P., Marshall, S.M., and Raymont, J.E.G. (1947) An experiment in marine fish cultivation. *Proceedings of the Royal Society of Edinburgh, Section B* 63 (1), 1–5.

Haime, J. (1874) The history of fish culture. In: *Report of the Commissioner for 1872 and 1873, Part 2*, pp. 465–492. U.S. Commission of Fish and Fisheries, Washington, D.C.

Hecht, T. and Britz, P.J. (1990) *Aquaculture in South Africa*. Aquaculture Association of South Africa, Pretoria.

Hepher, B. and Pruginin, Y. (1981) *Commercial Fish Farming with Special Reference to Fish Culture in Israel*. John Wiley & Sons, Inc., New York.

Hickling, C.F. (1971) *Fish Culture*. Faber and Faber, London.

Hines, N.O. (1976) *Fish of Rare Breeding*. Smithsonian Institution Press, Washington, D.C.

Hora, S.L. and Pillay, T.V.R. (1962) *Handbook on Fish Culture in the Indo-Pacific Region.* FB/T14, Food and Agriculture Organization of the United Nations, Rome.

Hornell, J. (1950) *Fishing in Many Waters.* Cambridge University Press, New York and London.

Houghton, W. (1895) *British Fresh-Water Fishes.* A. Brown & Sons Ltd., Hull.

Huet, M. (1972) *Textbook of Fish Culture.* Page Bros. (Norwich) Ltd., Norwich.

International Council for the Study of the Sea. (1903) *Report of Administration for the First Year.* The Royal Geographical Society (with the Institute of British Geographers), London.

Jhingran, V.G. (1975) *Fish and Fisheries of India.* Hindustan Publishing Corporation, Delhi.

Lever, C. (1996) *Naturalized Fishes of the World.* Academic Press Ltd., London.

Lin, S.Y. (1949) Notes on fish fry industry of China. *First Meeting of the Indo-Pacific Fisheries Council,* pp. 65–71. Food and Agriculture Organization of the United Nations, Rome.

MacDougall, E.B. (editor) (1987) Ancient Roman villa gardens. In: *The Tenth Colloquium on the History of Landscape Architecture.* Dumbarton Oaks, Trustees for Harvard University, Washington, D.C.

MacKenzie, C.L. (1996) History of oystering in the United States and Canada. *Marine Fisheries Review* 58, 4.

Matheson, C. (1929) *Wales and the Sea Fisheries.* National Museum of Wales, Cardiff.

Matsui, I. (1970) *Theory and Practice of Eel Culture.* Nihon Suisan Shigen Hokaku Kyokai Publishers, Tokyo [translated from Japanese].

Mills, D. (1971) *Salmon and Trout.* Oliver & Boyd, Edinburgh.

Netboy, A. (1974) *The Salmon: Their Fight for Survival.* Houghton Mifflin, Boston.

Newton, L. (editor) (1951) *Seaweed Utilisation.* Sampson Low, London.

Prince, E.E. (1906) *The Progress of Fish Culture in Canada. Sessional Paper No. 22,* pp. 89–107, Department of Marine and Fisheries, Ottawa, Canada.

Radcliffe, W. (1926) *Fishing from the Earliest Times.* John Murray, London.

Rivinus, E.F. and Youssef, E.M. (1992) *Spencer Baird of the Smithsonian.* Smithsonian Institution Press, Washington, D.C.

Russell, F.S. and Yonge, M. (1975) *The Seas.* Frederick Warne & Co., Ltd., London.

Schuster, W.H. (1949) *De viscultuur in de kustvijvers op Java (Fish culture in the coastal ponds of Java).* Publicatie No. 2, van de Onderafdeling Binnenvisserij. Vorkink, Bandung [in Dutch].

Shelbourne, J.E. (1964) The artificial propagation of marine fish. *Advances in Marine Biology* 2, 1–83.

Slack, J.H. (1872) *Practical Trout Culture.* Orange Judd & Co., New York.

Staikov, Y. (1994) *Development of Aquaculture in Bulgaria. Fisheries* 19 (2), 28.

Summers, C.C. (1964) *Hawaiian Fishponds.* Bernice, P. (ed.) Bishop Museum Special Publication No. 52, Bishop Museum Press, Honolulu.

Supino, F. (1913) Carp-breeding in rice fields in Italy. *Bulletin of Agricultural Intelligence and Plant Diseases* 4 (9), 1332–1335. International Institute of Agriculture, Rome.

Thomas, R.G. and Wilson, A. (1994) Water supply for Roman farms in Latium and South Etruria. *Papers of the British School at Rome* 62, 139–196.

Thompson, W.W. (1913) *The Sea Fisheries of the Cape Colony.* T. Maskew Miller, Cape Town and Pretoria.

Thorpe, J.E. (1980) *Salmon Ranching.* Academic Press, London.

Vibert, R. and Lagler, K.F. (1961) *Pêche Continentales (Continental Fishing).* Dunod, Paris [in French].

Wimpenny, R.S. (1963) The plaice. *The Buckland Lectures for 1949.* Edward Arnold & Co., London.

Wood, E.M. (1953) A century of American fish culture, 1853–1953. *Progressive Fish Culturist* 15(4), 147–163.

Worthington, S. and Worthington, E.B. (1933) *Inland Waters of Africa.* Macmillan & Co. Ltd., London.

Yarrell, W. (1941) *A History of British Fishes.* Van Voorst, London.

Journals, Society Publications, Magazines, and Trade Papers

Aquaculture Magazine (1979 onward) and all its consolidated predecessors from 1969 (*The Catfish Farmer, U.S. Trout News, American Fish Farmer, Commercial Fish Farmer,* and *World Aquaculture News*).

Bulletin of the U.S. Fish Commission (1881–1903) http://docs.lib.noaa.gov/rescue/Fish_Commission_Bulletins/data_rescue_fish_commission_bulletins.html.

Fish Farmer. (1978 onward). Special Publications (editor@specialpublications.co.uk).

Fish Farming International. (1974 onward). Arthur Heighway and EMAP Heighway. London.

Fisheries Bulletin. (1950–1957) Food and Agriculture Organization. Rome.

The Israeli Journal of Aquaculture—Bamidgeh. (1955 onward). Mires, D. (ed.) (ija-editor@siamb.org.il).

Rapports et Proces-Verbeaux des Reunions, Bulletin, et Journal du Conseil (1903 onward). International Council for the Exploration of the Sea.

Tilapia Varia. House Magazine of the Tilapia International Foundation, Utrecht (tif@tilapiastichting.nl).

U.S. Commission of Fish and Fisheries Annual Reports (1871–1903) *Bulletin of the U.S. Fish Commission.* Vols. 1–23. U.S. Government Printing Office. Washington, D.C. (http://penbay.org/cof/uscof.html).

Yamaha Fishery Journal. (1986) Composite Edition of Volumes 1–27. Yamaha Motor Company, Shizuoka-ken, Japan.

End Note

Dr. Colin Nash developed the first version of *A History of Aquaculture* in 2000. An illustrated lecture for the American Fisheries Society meeting followed in 2007 in San Francisco, California. Encouraged by the response from that presentation, Dr. Nash made the decision to write a book with the same title, leading to a first draft manuscript. Several rewrites and versions followed, leading to a second assembly of the manuscript in February 2008.

The final version was edited by Dr. Susan Thomas and submitted for publication to Wiley-Blackwell in March 2010. In his book, Dr. Nash draws from his distinguished international experience in aquaculture research and development to provide a unique and comprehensive historical narrative on the development of aquaculture around the world from ancient to modern times, and on the growing global importance of fish and shellfish as foods.

The Pacific Aquaculture Caucus, Inc. (PACAQUA) began assisting Dr. Nash in getting the book into print in March 2008 at the party celebrating his retirement from the National Oceanic and Atmospheric Administration. Development of a final manuscript of the book, including a new introductory chapter added by the author, *Fish and Shellfish as Food*, began in spring 2009. The new chapter was necessary to introduce the importance of fish and shellfish in the development of modern humans and their culture. With a final draft manuscript in hand, the author and PACAQUA sought a publisher, with the assistance of the United States Aquaculture Society (USAS), a chapter of the World Aquaculture Society (WAS). Wiley-Blackwell accepted the manuscript and requested a title change to *The History of Aquaculture*, given the distinctive perspective of the book.

PACAQUA is an aquaculture organization with a membership whose interests span the range of West Coast industry, academia, government agency, and individual members' aquaculture activities. Its mission is to promote economically viable and environmentally responsible marine and freshwater aquaculture for the Pacific region through sound public policy and best available science.

USAS is a professional organization dedicated to the exchange of information and networking among the diverse aquaculture constituents interested in the advancement of the aquaculture industry through the provision of services and professional development opportunities. Its mission is to provide a national forum for the exchange of timely information among aquaculture researchers, students, and industry members in the United States.

A goal of both organizations is the publication of aquaculture-related materials important to United States aquaculture development. This book project is possible because of a unique partnering of the author, PACAQUA, USAS, and Wiley-Blackwell Publishing. We wish to thank Dr. William Fairgrieve and Dr. Michael Rust of PACAQUA for their enthusiasm and perseverance to bring this book project to fruition, and Justin Jeffryes of Wiley-Blackwell for his cooperation. The USAS Publications Committee members include Drs. Wade Watanabe (Chair), Jeff Hinshaw, Jimmy Avery, and Christopher Kohler, with Douglas Drennan and Wendy Sealey serving as immediate past and current presidents, respectively.

—Peter Becker, Ph.D.
Chairman, Pacific Aquaculture Caucus, Inc.
Manchester, Washington
CEO, Olympic AquaFarms-BP/S Industries, Inc.
Port Angeles, Washington

—Wade O. Watanabe, Ph.D.
Director and Publications Chair, United States Aquaculture Society
Mariculture Program Leader, Marine Biotechnology in North Carolina
Research Professor and Aquaculture Program Coordinator,
Center for Marine Science, University of North Carolina
Wilmington, North Carolina

Index

Printed and bound by CPI Group (UK) Ltd, Croydon, CR0 4YY

Printed and bound by CPI Group (UK) Ltd, Croydon, CR0 4YY

16/04/2025

14658605-0002